Environmental Activism and the Media

This book is part of the Peter Lang Media and Communication list.
Every volume is peer reviewed and meets
the highest quality standards for content and production.

PETER LANG
New York • Bern • Berlin
Brussels • Vienna • Oxford • Warsaw

Maxine Newlands

Environmental Activism and the Media

The Politics of Protest

PETER LANG
New York • Bern • Berlin
Brussels • Vienna • Oxford • Warsaw

Library of Congress Cataloging-in-Publication Data

Names: Newlands, Maxine, author.
Title: Environmental activism and the media: the politics
of protest / Maxine Newlands.
Description: New York: Peter Lang, 2018.
Includes bibliographical references and index.
Identifiers: LCCN 2017038643 | ISBN 978-1-4331-3118-9 (hardback: alk. paper)
ISBN 978-1-4331-5010-4 (paperback: alk. paper)
ISBN 978-1-4331-5011-1 (ebook pdf) | ISBN 978-1-4331-5012-8 (epub)
ISBN 978-1-4331-5013-5 (mobi)
Subjects: LCSH: Mass media and the environment—Australia. | Mass media and
the environment—Great Britain. | Mass media and the environment—
United States. | Environmentalism—Australia. | Environmentalism—
Great Britain. | Environmentalism—United States.
Classification: LCC P96.E57 N49 | DDC 320.58—dc23
LC record available at https://lccn.loc.gov/2017038643
DOI 10.3726/b13241

Bibliographic information published by **Die Deutsche Nationalbibliothek**.
Die Deutsche Nationalbibliothek lists this publication in the "Deutsche
Nationalbibliografie"; detailed bibliographic data are available
on the Internet at http://dnb.d-nb.de/.

For Stephen and Mum & Dad

CONTENTS

ILLUSTRATIONS

ACKNOWLEDGEMENTS

I would like to acknowledge some of the people whose contribution, ideas and insights have gone into this book. Thanks to Kathryn Harrison, Mary Savigar, Michael Doub and all the team at Peter Lang Publishing. Thanks to the Mirrorpix and *London Evening Standard* for the permission to reproduce the newspaper images. This work has benefited from the activists and individuals who have given up their time to talk and be interviewed over several years. Special thanks goes to John Jordan, Dan Glass, John Stewart, Martin, Mike, Steve, Debra, Nim, Des and London Critical Mass, The Hive, London Action Resource Centre (L.A.R.C.), VisionOnTv, Occupy LSX, new activists friends, and new colleagues at James Cook University in Australia and all the other activists who have broadened my knowledge – your experience and insights are at the heart of this book.

This book owes an immeasurable debt of gratitude to Professor Erika Cudworth and Dr Stephen Hobden, as both my PhD supervisors, mentors and friends. Thanks also to Dr Abel Ugba, Professor Libby Lester and Professor Heather Nunn for reading through earlier versions and providing support and guidance. Thanks to the reviewers for their time and suggestions. Dr Angie Voela, thanks for your support, some of chapter seven is taken from our work together. Thanks too to the E16 cocktail club. I also would like to thank Professor Iain MacRury and Dr Andrew Calcutt for their guidance and editorial help with chapter three. Chapter Three is partially taken from an edited collection by Iain, Andrew and Professor Gavin Poynter titled *London After Recession A Fictitious Capital?* (2012) Routledge.

Lastly, but never least, my wholehearted thanks to my family and tribe. To Anna, Nick, Otto and Teddy, Si Cobb, Mark and Kazza, Joey, Simon and the Parsons tribe—thanks for asking when writing was going well, and staying quiet through the difficult times. My family, Mike, Uncle Alan and Aunty Maureen, and Bill and Ann Newlands, your quiet encouragement has been a constant source of support. For my parents—Mum you carry yours and dad's spirit with you in all your love and support-thank you.

To Stephen, I dedicate this book to you for all your love, support and the many late night talks that makes life with you so adventurous.

ABBREVIATIONS

ACPO	Association of Chief Police Officers
BAA	British Airports Authority
BRICS	Brazil, Russia, India, China and South Africa
BP	British (Beyond) Petroleum
CDA	Critical Discourse Analysis
CFCs	Chlorofluorocarbons
CJA	Criminal Justice Act
CJB	Criminal Justice and Public Order Bill
CND	Campaign for Nuclear Disarmament
CoP	Conference of the Parties
CSC	Counterspin Collective
CSD	United Nations Commission on Sustainable Development
DEFRA	Department for Environment, Food and Rural Affairs
DoE	Department of Environment
EMSs	Environmental Management Systems
EMT	Ecological Modernization Theory
ENGOs	Environmental Non-governmental Organizations
EU ETS	European Union Emissions Trading Scheme
GATT	General Agreement on Tariffs and Trade

GEG	Global Environmental Governance
HACAN	Heathrow Association for the Control of Aircraft Noise
IBRD	International Bank for Reconstruction and Development
IMF	International Monetary Fund
IPCC	Intergovernmental Panel on Climate Change
KACAN	Kew Association for the Control of Aircraft Noise
LARC	London Action Resource Centre
MAFF	Ministry of Agriculture, Fisheries and Food (now DEFRA)
MEAs	Multilateral environmental agreements
MoD	Ministry of Defence
NAFTA	North American Free Trade Agreement
NECTU	National Extremism Tactical Coordination Unit
NEPI	New Environmental Policy Instruments
NFU	National Farmers Union
NOTRAG	No Third Runway Action Group (Heathrow Airport)
NPOIU	National Police Intelligence Unit
PCC	Press Complaints Commission
PPPs	Public and Private Partnerships
RIPA	Regulation of Investigatory Powers Act
SMOs	Social Movement Organizations
SOCPA	Serious Organized Crime and Police Act (2005)
TAZ	Temporary Autonomous Zone
TNR	Trans Northern Companies
UN	United Nations
UNCED	United Nations Conference on Environment and Development
UNEP	United Nations Environmental Programme
UNFCCC	United Nations Framework on Conventions on Climate Change
WTO	World Trade Organization

· 1 ·

INTRODUCTION

In 1999, Timothy Luke wrote that, "in the ongoing struggle over economic competitiveness, environmental resistance can even be recast as a type of civil disobedience, which endangers national security, expresses unpatriotic sentiments, or embodies treasonous acts" (1999, p. 125). Luke was observing what would become a political shift in the role of environmental activism in Western Democracy. In the same year the global justice movement exploded onto the world's stage with anti-capitalist protests in Seattle, USA (1999). Protesters gathered to bring public attention to the environmental impact from trilateral agreements taking place at the World Trade Organization meeting that would have long-term economic and environmental impacts on most people. The ensuing battle between these two groups – political economists and environmental activists has produced binary positions in seeking solutions to climate change. In the fight, some environmental activism and protest more generally has become criminalize in the Western World, whilst economists and liberal politicians push for more and more market solutions. This book is based upon a critical examination of the media and political representation of environmental activism in shaping public knowledge about climate change. The book is based on interviews with environmental activists and critical discourse analysis of newspaper representations of environmental activism from the 1980s to the early 2010s.

Environmental Activism, Politics and Traditional Media[1]

Over the past four decades environmentalism has become the baseline for key areas of political, economic and social justice debates. Growth within the professional political left of Environmental Non-Governmental Organizations (ENGO) has seen an expansion of interest groups such as Greenpeace, Sea Shepherd, Friends of the Earth, World Wildlife Foundation etc., into multinational organizations. Equally, radical environmental activists have seen different incarnations like the Global Justice Movement, Climate Camps, Occupy and Anti-Fracking Movements. Simultaneously, global Institutions and political collaborations has led to mainly non-binding agreements with the creation of United Nations Framework Convention on Climate Change and Agenda 21 (1992), the Kyoto Protocol (1997), Bruntland agreement, Conference of Parties (CoP) (1992–present) including the Paris Agreement (2015) and many more. Each of these initiatives places economics at their core. The result is a split between governments and global institutions seeking economic led solutions and the more sociological solutions of the left and radical left, as evidenced by the continuation of environmental protest around the planet despite moves to criminalize and curtail.

Across the world, environmental activists are constantly challenging neoliberal approaches such as energy markets, emissions and carbon trading schemes that has become common as a solution to climate change. How activism movements maintain pressure on governments and corporations is the focus of the second half of this book. From social media to superglue, artivism, craftivism and spy cams, activists are constantly developing innovative ways to be a voice within environmental politics. This book focuses on the radical environmental activists collectives such as the Anti-roads movement and Reclaim the Streets (1990s), Global Justice Movement (2000s) mega-protests at Conference of Parties (CoP), Camps for Climate Action UK (2005–2010), Reclaim the Power (2008-present), (2010–), Occupy Movement (2011) and anti-fracking—Knitting Nanna's (2012) and the Red Lines (2015–2017).

To distinguish between professional and radical, hierarchical and non-hierarchical this book refers to radical as ecoActivists/ism with a capital A, and ecoActivism for professional, hierarchical organizations. The A represent the inclusion of the anarchist strand in radical protest collectives such as Black Bloc (1990s–present). Black Bloc is not a prescribed collective, but more a group of individuals who believe in "violence against the police as a legitimate political tool, a form of self-defence against the state" (Viejo, 2003, p. 371). Much of the Black Bloc rhetoric came out of the Class War, Poll Tax Riots (1989) and Whitechapel Anarchist collectives (1980 onwards), and still permeates the movement today. EcoActivists from each of the groups identified above, and other collectives were interviewed for this book. The book draws on over 50 interviews with activists in the UK, USA and Australia. Critical Discourse Analysis of newspaper clippings and reporting of ecoActivism in the traditional media as the framework for the findings and supported by interview data and focus groups. Drawing on media (Fairclough, 1995; Couldry, 2000; Gerbaudo, 2012) and political discourse theory (Dryzek, 2000; Foucault, 1977, 1991), this book charts the relations of power and socio-economic outcomes between ecoActivists and mainstream polity since 1997. In doing so the power relations reveal that shaping the public understanding of radical environmental activism sees power shifting away from activism towards environmental governance and free-market economics, nestled in a media discourse. Climate change debates and the political economy have become so entrenched in media practices we are creeping towards deadlock. This book applies a Foucauldian interpretation of his concept of governmentality with Luke's green governmentality to show that neoliberal politics takes us beyond Luke's prediction, recasting anyone from an individual to international ENGOs within a frame of deviance.

Governmentality and Green Governmentality: The Acceptable and Unacceptable Environmentalist

The guiding theories in this book draw on Foucault's concept of discourse, power, biopower, governmentality and heterotopia (1971, 1977, 1982, and 1991). The book takes Luke's interpretation of governmentality (see Chapter Three) and green governmentality (Luke, 1999) to examine the triangulation of power in shaping public opinion and knowledge on climate change. Green governmentality, or what Oels (2005) terms eco-governmentality, helps chart shifts in power within environmental discourse.

Green governmentality draws on Foucault's idea of governmentality through a green lens and the notion of biopower. Foucault shows that governments can govern in either way that rewards people who side with their policies and ideologies (known as technologies of the self) or discipline (known as technologies of dominance) those who contest those who challenge the policies or ideologies. This bifurcation of power is unpacked with the Foucauldian notions of biopower and biopolitics—the body as a political tool.

Thus climate change solutions can be seen through the lens of these two technologies. Individuals and groups who agree that economics is the best solution are rewarded with green points from supermarkets, or a sense of public duty, of saving the planet. Whereas individuals who reject capitalists' solutions, in favour of a pluralistic approach, such as ecoActivist collectives, can often be excluded from the debate. As Chapter Four will show, ecoActivists and individuals are often subject to state-sponsored surveillance and laws that curtail protest as a technology of dominance.

Drawing on Foucault's notion of governmentality this book explores emerging political discourse that position the climate change debate between people who support economic solutions, over those who want a more egalitarian approach that includes climate justice. In understanding governmentality, this book will firstly traverse the world of governance, geopolitics, biopolitics, eco-governance and green capitalism (Newell and Paterson, 2010) before moving onto a series of case studies to discuss the consequences of such greening of the state. The case studies show how environmentalism has taken two different routes, economics and pluralistic as a "key framing and mobilising discourse [that] articulated a new political agenda for mobilising climate activism" (Chatterton, Featherstone, and Routledge, 2013, p. 603).

Thus, governmentality can be a concept that both facilitates a way to describe a neoliberal or advanced liberal rule, and also as a framework or method that examines "mentalities or rationalities of government" (Death, 2010). Governmentality can be a specific form of power, held by the state over the populations. Governmentality can also be a way of analysing relations of power in general. In this book governmentality is applied as a framework for analysing relations of power between governments, traditional media and environmental activists.

As we'll see, Western governments use techniques of dominance to pass new police powers of arrest and detainment against people who challenge economic led environmental solutions through protest. The political art of environmental activism has become to be seen within a discourse of an accepted threat to state security. This book sets out how political and media framing of radical environmental activism is increasingly wrapped in a discourse as a threat to societal order and security as justification for state sponsored surveillance and criminalization of various places and forms of protest. This book is based upon a critical examination of the consequences for environmental activism as a democratic right in a world that favours economics over social, egalitarian solutions to climate change.

To understand the role of green governmentality in the reporting of ecoActivism this book draws on Critical Discourse Analysis (CDA) for its analysis and original interview data from a range of activists. CDA is preferred over text or content analysis as it provides analytical tools that contextualize theoretical positions. Contextualization of a news story is necessary for frames to have maps of meaning. Contextualizing comes from the "methodological field of history" (Foucault, 2008, p. 12). Without contextualization, a back-story to use journalistic jargon, any story struggles to makes sense. Like the climate debate discussion in Chapter One, in order to understand traditional media's representation of environmental activism we need to contextualize the events by looking at previous media reports. Media discourse analysis helps unpack how sentence structures (Van Dijk, 1988), framing (Gamson and Modigliani, 1989) and linguistic traits can produce discourse and meaning. In addition to this, CDA provides the intertextual tools to examine both the mechanics of meaning and the theoretical context in order to address the research questions. Therefore, Fairclough's Critical Discourse Analysis (1995) is the preferred empirical research tool for analysing newspaper reports.

Traditional media (television, radio and print) and new media (social and online) practices play a central role in this debate. Traditional media as the fourth estate's role is to decipher and challenge the two sides of the argument, economic (neoliberal) and egalitarian (activists) for the public. The first environmental journalist was appointed by the *New York Times* newspaper in the mid-1960s signalling a recognition that the environment was becoming a social and media issue (Lester, 2010). Reporter for the *New York Times* newspaper, Gladwyn Hill is credited with being one of America's first national environmental correspondents. A. M. Rosenthal, former executive editor and columnist of the *New York Times* says, "Gladwin Hill helped open up a whole new field of reporting, and became enormously important to the country and to journalism" (Perez-Pena, 1992). Adding that as a columnist, "he had tremendous journalistic skills, and a deep interest in the environment long before it was fashionable" (Perez-Pena, 1992). Hill and other journalists around the world were tapping into and exposing an emerging political and social understanding of the environment.

Today, new or social media enables environmental groups to circumnavigate traditional media practices and create their own discourse through media stunts and spectacles. Media image events (Deluca, 1999) have become common throughout the last five decades of radical ecoActivism. Memorable media image events include women in the 1980—some women intentionally goaded police into arresting them, to generate an image event of a mature lady being carried from Greenham Common by police officers (1983). The late 1990s anti-roads activist Swampy emerging into the media spotlight (1980s) as a media event, and the Reclaim the Streets (RTS) parties (1980s) all generated image events for the waiting press. Around the world, Tar Sands protests in Canada (2006); Global Justice Movement protests (1990–2000), repeated with the Occupy and Arab Spring movements (2011) and the eponymous Battle for Seattle (1999) as masked-up protesters mingled with people dressed as turtles and a series of image events of the global justice movement. When Australia hosted the G20 meeting (Brisbane, 2014) a new form of artivism protest saw a 100 people all with their Heads in the Sands of Bondi Beach, near Sydney Australia (2014). At the UNESCO Conference of Parties (CoP) 21 in Paris, artivism saw giant inflatable cobblestones blocking roads. Also at CoP 21 South American and Pacific Island Nations indigenous peoples were marching against the developed worlds for justice (2015)—each event an image and snapshot of the environmental movement. Early environmental journalism studies shows a "clear and widespread media discovery of the 'environment' in 1969" and there was a clear indication that "international flows were clearly at work in the formation of the environment as a social problem and media issue" (Lester, 2010, p. 32). Today journalists and politicians tend to refer to these changes as climate change, global warming and the Anthropocene.

Climate Change Concepts: A Potted History

The Anthropocene, global warming, climate change, environmentalism and acid rain are all terms referring to the impact of a high carbon society on the earth's weather systems. To be clear, global warming is increases in overall global temperatures. Climate change includes increasing temperature, but also represents the changes in the planet such as sea level rises, melting glaciers, ocean warming causing ocean acidification, dramatic changes in weather patterns leading to increase in flooding and drought. Buckingham and Turner (2008) say that climate change is "any change in the global climate system, over time, whether due to natural variability or as a result of human activity" (2008, p. 179). Whereas global warming is the "recent climatic amelioration, observed across the globe, believed to be a result of anthropogenic (human) forcing due to increased release of greenhouse gasses into the atmosphere" (2008, p. 179). Environmentalism is the interaction between people and planet in a socio-political context affected by social, political and cultural relations. Another way of looking at it is climate change and global warming often refers to scientific based terminology; environmentalism is a social science term. Environmental activists focus on the social and rely on the science to prove the argument.

Evidence suggests that anthropogenic (human relations) impacts on nature first began in the latter part of the nineteenth century. The industrial period was the start of a high carbon economy that was to grow exponentially over the next two centuries. A postwar economic boom (1967–1997), a dominant American Imperialism model pushing for higher consumption and ramping up export and imports, including road vehicles, aeroplanes and infrastructure adds to the high carbon society. The end of the twentieth century was an era epitomized by consumerism and capitalism cosying-up with neoliberal politics to drive the global economy. The new millennium sees an increased capitalism driving a higher carbon economy as mass production and population from developing and developed countries (i.e. Asia and Indo Pacific regions) feeds the global north's appetite for consumerism putting more and more pressure on the environment.

An Emerging Anthropocene?

Questions over whether a high carbon economy could affect the atmosphere began around the late nineteenth century (1880). The period from 1880 to the present day is a time frame that many scientists are referring to as the Anthropocene. The Anthropocene is a term deriving from geological studies to describe the "central role of mankind in geology and ecology by proposing to use the term 'Anthropocene' for the current geological epoch" (Crutzen and Stoermer, 2000, p. 41). British Chemist Robert Angus Smith's (1872) first recorded study was testing if pollution (carbon from coal burning, chemicals in wool production) from the textile mills of northern England were affecting local weather patterns. Smith's findings show that rainfall in the Greater Manchester area had an effect in the levels of acid in the air, which led to him coining the term "acid rain". Smith also "discovered a link between acid in rain failing near Manchester and the sulphur dioxide given off when coal is burned. Seventy years later, research in Europe confirmed this discovery (…) thus the phrase acid rain entered the language" (McCormick, 1989, p. 5). Smith's "acid rain" becomes an allegory for the impacts of industrialization on the global climate for the next 100 years.

Not until the mid 1970s did scientist Wallace S. Broecker (1975) define the parameters of what we understand by global warming and climate change. Broecker's paper *Climatic Change: Are We on the Brink of a Pronounced Global Warming?*[2] began the debate on what we mean by global warming and how it differs from the term climate change. Broecker observes that "by the first decade of the next century we may experience global temperatures warmer than any in the last 1000 years" (1975, p. 461). Researching Greenland's glaciers, Broecker (1975) questions the impacts of man-made chemicals has on the global temperatures and finds that "as the CO_2 effect will dominate, the uncertainty here lies mainly in the estimates of future chemical fuel use and in the magnitude of the warming per unit of excess atmospheric O_2" (1975, p. 463). Adding, that "without regulation to monitoring global temperatures the earth is in for a climatic surprise" (1975, p. 463). Broecker points out that the post-war period has seen global temperatures adjust through natural balance as "the major point of the argument is that over the past 30 years the warming trend due to CO_2 has been more than countered by a natural cooling" (1975, p. 463). He warns that natural weather patterns and cycles cannot continue to compensate for the Anthropocene, and "this compensation cannot long continue both because of the rapid growth of the CO_2 effect and because the natural cooling will almost certainly soon bottom out. We may be in for a climatic surprise" (1975, p. 463) and warns that without drastic change that:

> The onset of CO_2 induced warming may be much more dramatic than in the absence of natural climatic variations. There is little doubt, however, that this gradual warming will lead to changes in the pattern of global precipitation. Our efforts to understand and eventually to predict these changes must be redoubled. (1975, p. 463)

Smith (1872), Broecker (1975) and Crutzen and Stoermer (2000) studies began over concerns about the environmental consequences of industrial revolution. We see the trajectory of terms from acid rain to global warming, climate change and today the Anthropocene. The two terms have become interchangeable in the media, and appropriated by politicians, journalists and academics as a catch all phrase for environmentalism. To be clear, climate change is according to the science, the heating up of the global temperatures due to human activity. Global warming is the long term warming of the planet over time that is often referred to as the Anthropocene. The Anthropocene may have begun in the 1880s, but only very recently has the human race come to accept that high carbon economies have a recognized impact on the global temperature. Not until 2013, over 141 years after Smith's (1872) first study, was there a formal recognition and 95% certainty that the climate is changing due to human influence. The United Nations Intergovernmental Panel on Climate Change (IPCC) fifth report (Stocker, Field, Edenhofer, 2014) notes due to "human influence has been detected in warming of the atmosphere and the ocean, in changes in the global water cycle, in reductions in snow and ice, in global mean sea level rise, and in changes in some climate extremes" (2013, p. 13). Whilst each term has separate meanings, for ecoActivists, ENGO and the environmental left the unifying discourse is the impact high carbon economies have on the environment. Since the late 1960 and the Earthrise image environmental concerns emerging with a focus on the human impact to the planet from industrialization and rapid increases in consumerism.

The Rise of the Environmental Left

Today, ENSM sit as the formal, hierarchical organizational structures (old social movements) that engage with political institutions in advancing political integration and economic rights (Martell, 1994). Old social environmental movements include Oxfam, Sierra Club and World Wildlife Fund. Environmental New Social Movements (ENSM) began gaining traction in the 1970s and early 1980s (Anderson, 1997), such as Greenpeace, Friends of the Earth, 350.org and Sea Shepherd. Emerging in an age that saw increasing politicization of the environment, ENSMs have become a major part of environmental politics all across the globe. Expansion in mass media, youth culture, social movements, civil rights, feminism, ecofeminism, technology, the internet, World Wide Web (WWW) has been adopted by both professional and radical environmental movements as we witness large-scale bottom-up protests and civil disobedience.

Environmental activism within the politics of protest emerges around the same time as new social movements (NSM) in the late 1960s. A movement is defined as people with a desire for a "sort of radically democratic political economic and cultural transformations, and who believed in undertaking some sort of non-violent direct action" (Gitlin, 1980, p. 293). There are two strands of NSM—old and new (Martell, 1994). Old social movement practices include a top-down hierarchical staff organization found in any multinational, they are rich in resources and have global reach. New social movements often defined through the more radical direct action and civil disobedience focuses on social justice and single issue political participation. NSMs are more in line with identity politics and "raise issues and make demands about identity, lifestyle and *difference*" (Martin, 2015, p. 61).

Within ENSM there are two strands—hierarchical and non-hierarchical organizational structures, or professional and radical. Radical ENSM are movements emerging as autonomous civil society collectives of coordinated informal networks and grassroots activism, non-violent direct action and cultural politics. Radical is a collective of likeminded group of people—leaderless and all decision-making processes are based on collective consensus. Anderson (2004) describes such collectives as made up of individuals at a grass-roots level, which "operate under a variety of organizational (and disorganization) banners depending on the action concerned" (2004, p. 107); or the non-hierarchical, horizontal collectives of like-minded individuals that form together either ontologically to create new activist movements, or as part of the wider environmental activist movement as a whole.

David Foreman (2005) co-founder of the radical Earth First! sees professional or old ENGO models as "political movements become more debating societies where the participants engage in philosophical masturbation and never get down to the vital business at hand" (Foreman, 2005, p. 352). Forman describes radical collectives with values as "none of us will even transcend a polluting lifestyle—but we can act. We can act with courage, with determination, with love for things wild and free. We can't be perfect, but we can act. We are warriors. EarthFirst! is a warrior society. We have a job to do" (Foreman, in Dryzek and Schlosberg, 2005, p. 352). The simplest definition comes from MacKay in his book *DiY Culture, Party and Protest in Nineties Britain* "activism means action" (1998, p. 5).

Managing the Environment and Wilderness

Parallel to emerging environmental social justice movements was the formalization of global green political discourse (1970s; Torgerson, 1999). In the USA, the first Earth Day 1970 raises awareness of high carbon economy on the global environment. The day was later formally adopted by the United Nations Educational, Scientific and Cultural Organization (UNESCO). Political policies and processes in managing the environment see governments in North America, the UK and Australia create departments of the environment, Clean Air Acts and Water management procedures. Such managerialism of nature was first seen centuries earlier, through the national park policies.

Before the 1970s, environmentalism, particularly in the UK and North America, often hung on questions on consequential impacts of industrialization, mass food production, conservation and the nuclear debate. A couple of decades before, and the Campaign for Nuclear Disarmament (CND) were regularly organizing protest marches between the Atomic Weapons Research Establishment in Berkshire into central London. In the USA, Rachel Carson's seminal book *Silent Spring* (1962) was raising questions on pesticide usage in mass food production. Then in 1968, the Earth Rise image draws together the separate debates and the birth of the environmental movement.

Earthrise—our blue planet in the darkness of space—became eponymous with people's perception of Earth, nature and our relationship to the planet (Dryzek, 1997; Doyle, 2007; Gore, 2007; Lester, 2010). America's National Aeronautics and Space Administration (NASA) Astronauts Borman, Lovell and Anders aboard the Apollo eight mission broadcast live images of the earth as seen from lunar orbit. The image – Earthrise – triggered a new wave of environmentalism that moved away from conservation and towards environmental justice. Academically, politically and culturally Western societies take the images and begin examining green political discourse as the "intersection of ecosystems and human social systems" (Dryzek, 1997, p. 8).

The iconic image has today become synonymous with the environment and used in advertising, branding logos, public relations and the mainstream media. Environmental groups use the symbolism of the global as an emblem (i.e. Earth Hour (March 28th, 2007–present), and Earth Day (April 22, 1971—present)). Media organizations incorporate the globe (CNN, BBC and ABC) into their branding. Wikileaks has two globes in an hour glass, Anonymous logo is a man (albeit with a question mark for a head) in front of a culture jammed UN global logo. The United Nations (UN) logo is a globe, the Fédération Internationale de Football Association (FIFA) is a globe and so on.

Cottle and Lester (2011) observe "depictions of the globe in the context of news on climate change have also now become commonplace in the media, as they have in the iconography of new journalism more generally" (2011, p. 926). The Earth as an image plays a "prominent part in bringing home the threat and reality of global climate change, as well as positioning for many, particularly within the West, as possibility *the* crisis of the global age" (2011, p. 926). Whilst Earthrise instigated a shift in the environmental movement, it also began to change how politicians and the media report on environment and related socio-political issues and contestations.

The Rise of the Environmental Right: Climate Sceptics and the Conservative Right

Capitalism and neoliberal policies become the dominant discourse in the 1980s and early 1990s. UK Prime Minister Margret Thatcher's opened up new trade routes between Britain and mainland Europe, combined with deregulation of banking markets allowing for rapid growth, and new home ownership initiatives were all ways of removing state responsibility for social issues. USA President Ronald Reagan took similar steps to open up trade with the North America Free Trade Agreement. Reacting to Thatcher's new economic policies in the 1990s activists began mobilizing with frequent protests over class, privilege and economic policies. The Poll Tax riots at the end of the decade summarized ten years of unrest. Thatcher's Roads to Prosperity paper (1989) opened up trade routes between the UK and Europe. Trade expansion led to many roads protests from Twyford Down, M11/Wanstead, Newbury Bypass and the trope of Swampy emerging out of a tunnel at the A30 Devon bypass protest camp. In the cities the Rave scene was the start of a DiY protest (McKay, 1998) and emergent ecoActivism reclaiming the streets and society from the neoliberal conservative right.

In Australia, Conservative Prime Minister John Howard announced road plans through the Daintree Wet Tropical rainforest in far north Queensland to open up the route to tourism. The Daintree became a famous protest against land clearance for property development as "real estate interests threatened to destroy what remained of the virgin lowland rainforest" (Cohen, 1996, p. 82). Protesters fresh from action in the southern state of Tasmania travelled north in support of local activists. The Queensland government had on the advice of the UNESCO World Heritage Committee set aside land on the tip for the Cape Tribulation National Park; however, activists were worried that a planned "building of a connecting road through the national park from Cape Tribulation north to Bloomfield, would rip up 33 kilometres of lowland forest wilderness" (Cohen, 1996, p. 82). Activists came together to form the Cairns and far north Environment Centre as a base to protect the rainforest and adjacent Great Barrier Reef. There were concerns that the topsoil from the road would "result in land destabilisation, resulting from landslide scars and topsoil entering the reef environment" (Cohen, 1996, p. 82).

Just a decade later and the relationship between ecoActivists and the domestic government intensifies over the space of climate change knowledge. Increasing economization of the environment is scaffolded against an increase neoliberal discourse and capitalism-led solutions to climate change combined with new legislation that limited some protest. Additionally, state-sponsored undercover policing and intelligence gathered rationalized as security measures meant protesters and individuals ran the risk of arrest and/or being put onto a domestic extremists watch list, and no-fly lists.

Those wanting to go beyond the professionalization of protest and political participation through lobbying, letter writing, clicktivism or slacktivism and towards more non-violent direct action and civil disobedience could either have intelligence gathered on them or be demonized as domestic extremists, eco-terrorist or be arrested under public order offences. From 1999 onwards a series of new UK laws and police units were created that impacted on the ability to protest. The National Public Order Intelligence Unit (1999); Terrorism Act 2000. Part five; National Extremism Tactical Coordination Unit (2004); Serious Organized Crime and Police Act (2005); National Domestic Extremism Team (2005), Counter Terrorism Command (CTC, 2006) and the Police Reform and Social Responsibility Act (2011[3]) all of these have been used to gather intelligence and convict ecoActivist in the UK (see Chapter Four).

This book charts the trajectory of increasing criminalization of non-violent direct mobilizations that appears in newspaper reporting. As Chapter Two shows over the last few decades media reporting within a security narrative pervades through the reporting of ecoActivism. The result was a temporary loss of voice for many of the movements, blanked out of the discourse to be replaced by economic arguments as the solution to climate change, whilst state actors influence public opinion and political decision-making processes with economics at the heart of the debate.

UK Political Right and Co-opting the Environment

Green governance became a political hot potato at the 2006 local elections, when the Conservative Party put the environment at the heart of their campaign. The co-opting of environmentalism and climate change concerns was outwardly seen as a U-turn by the Conservative Party who traditionally favoured economic growth over other factions such as the environment. Similarly, New Labour Party drafted a Climate Change Bill and commissioned Lord Stern of Brentwood to find out the economic impacts of climate change (2006). The resulting Stern report, by former head of the World Bank Nicholas Stern was a risk analysis testing the impact of climate change on the UK.

Global recognition of climate change was reiterated with the UN IPCC fourth report. The IPCC AR4 (2007) notes that it is extremely likely that "human influence has been the dominant cause of the observed warming since the mid-20th century" (IPCC 5th Report into Climate Change). The fifth (2013) UN IPCC report is the first global report that comprehensively accepts the Anthropocene is real caused by a high carbon anthropogenic society. The report (10.3–10.6, 10.9) clearly attributes changes in the earth's temperature as due to human activity. The statement reads that "It is extremely likely that more than half of the observed increase in global average surface temperature from 1951 to 2010 was caused by the anthropogenic increase in greenhouse gas concentrations and other anthropogenic forcing together." This book agrees with the IPCC AR5 position.

Public perception of environmental and climate change issues has waxed and waned over the decades often influenced by the level of public concern and engagement. Cottle and Lester (2011) found that in the early days, climate change was little discussed as a scientific or social discourse in the traditional media. Carvalho notes climate change was initially low level in science stories and news reports (2007). As the scientific evidence grew, the first global agreement that high carbon impacted on the planet's ecosystems was the UN IPCC fourth and fifth reports (AR4/AR5) and a series of celebrity events. By 2007, the two reports and wider concerns led to calls "for urgent global mobilizations at the G8 Summit meeting, and then the UN-sponsored conference in Bali late that year … proved to be a transformative moment in the news career of climate change" (Cottle, 2009, p. 506).

Saffron, Hywell, Kurz, Wiersma, and Boykoff's (2015) comparative study between the UK and USA found the UK gave far greater coverage to the fifth IPCC report than the US. Saffron et al found that "the production and consumption of climate change media is a major constituent of the "cultural politics of climate change. For over a decade, studies have explored the ideologies and journalistic norms apparent in print media coverage" (2015, p. 380). The results show that in traditional media,

> the IPCC gained far more attention in the UK (87 articles, 20 broadcasts) than the US (30 articles, 6 broadcasts). Even considering the unequal broadcast lengths (some were 30 min, others an hour), UK broadcasters spent nearly five times more airtime reporting the IPCC than US broadcasters (1 h 23 min 53 s in the UK, 17 min 53 sin the US); a pattern also evident in print (67,385 words in the UK; 25,482 words in the US). (Saffron *et al.*, 2015, p. 381)

However, IPCC governing principles do not explicitly address the communication of sciences, and it would be tardy in a social media and digital culture age not to attempt to communicate their findings via the internet and social media. Activists and *Guardian* newspaper journalist Leo Hickman observes how

> If the IPCC had been far more open, transparent and responsive during that period then it might have been able to avoid, or at least dilute, much of this negative examination. More importantly, perhaps, it might have been able … to communicate the science better to a wider variety of audiences. That's not to say the IPCC have no communication strategy, and the strategy in place should be updated to "reflect the online communications environment it now operates within and seeks to influence". (2015, p. 285)

Furthermore, Hickman (2015) believes the IPCC need to engage a dialogue via social media and train staff in best practice for Twitter and other social media platforms. In today's multimedia world "It is not enough to publish IPCC reports online and then sit back and expect the wider world to read them in their entirety" (Hickman, 2015). The Twittersphere was prominent in Saffron *et al.* (2015) study of the IPCC Fifth Assessment Report (AR5). Their study found Twitter users (26,623), sent 64,219 tweets over a peak period of two to three days following the releases of the first working group (WG) report. The tweets carried a strong emphasis on what Saffron *et al.* (2015) categorize as "settled science" (SS) discourse as "an emphasis on the science of climate change (across any WG) and the broad expert consensus. there is considerable evidence of the need for action. Science has spoken, others must act. Uncertainty or scepticism quashed" (2015, p. 381).

Saffron's *et al.* (2015)[4] study offers some interesting areas of research and the inclusion on social media in a comparative study between two countries heavily linked to the climate change debates provides some new perspectives. However, it's not clear who out of the 26,623 users are climate scientists, environmental activists, NGO organizations, climate skeptics and/or deniers, journalists and politically linked users and who is the general public that the IPCC is often accused of neglecting the wider public. The most popular hashtags were #climatechange, #globalwarming being identified over the #IPCC and #AR5, then it's not certain they relate to the Fifth IPCC Assessment, and as discussed in this book, the language and discourse around these two terms is somewhat contested and adopted by a range of climate change advocates, sceptics and deniers. Social media also provides, to some extent, a way for grassroots, radical and NGO environmental collectives and organizations to be able to contribute to the public knowledge.

Who are Atypical EcoActivists?

Saunders, Grasso, Olcese, Rainsford, and Rootes (2012) identify four different types of environmental activists, 1) Stalwarts, 2) Novices; 3) Returners and 4) Repeaters. Stalwarts' and Returners tend to be male and unemployed, having been on the scene, on and off every year, they seek information from listservs, emails, meetings, collective, social spaces, squats and fellow activists. Novices and Returners tend to be employed and gather information from traditional and social media. Novices and Returners are less likely to be embedded in protest networks and their internal channels of communication. Novices have never participated in protest (before the one at which they were surveyed); Returners exhibit moderate intensity and commitment, but greater persistence.

Political alignment for novices and returners (2012) tends to be "further right on the left-right continuum", and stalwarts and "repeaters being most likely to hold left-libertarian views" (Saunders *et al.*, 2012, p. 15). The study also found that long-term stalwarts are jaded with democracy. This research found that many of today's stalwarts are in their late 40s onwards, and have been involved in protests since children at Greenham Common Peace Camp, the Poll Tax Riot of the 1980s to the global justice and UNFCCC CoP summits. As Chapter Four will discuss, many of these stalwarts are witness to the increasing criminalization of ecoActivism. They have often been witness to or had suspicions (later proven) of undercover police within the movement, which goes some way in explaining their disillusion with democracy. Repeaters and returners (Saunders *et al.* 2012) "are generally more satisfied with democracy than their more engaged counterparts, and have participated in fewer political acts" (Saunders *et al.*, 2012, p. 2). The interesting people are those in the middle, between stalwart and novices.

Saunders *et al.* (2012, p. 2) identify 1) what sustains protest? and 2) protesters who are neither novices nor stalwarts are the implicit but unspecified reference group in studies that suggest "committed or persistent activists are more remarkable (structurally and/or in terms of engagement) than others". Stalwarts share knowledge and pass on techniques learnt from the anti-roads and Reclaim the Streets collectives of the 1990s. Additionally, they adapt traditional protest skills into a digital literate world, by running online, smartphone and digital journalism workshops, whilst others stick with traditional physical techniques through training camps and workshops on how they protest at major summit protests. As Plows (2006) notes stalwarts and repeaters act as mentors sharing information, techniques on protest camps setup and techniques to novices and returners. Saunders's *et al.* (2012, p. 19) found the group between the stalwarts and novices that keep protest movements going, and "often led by Stalwarts, who exhibit high levels of intensity and persistence" (2012, p. 19). In doing so the movements remain alive, if not dormant at times reshaping and reinventing the knowledge and praxis that enables them to start up again when necessary.

Academics, governments, police, activist's collectives and journalists are most interested in those that fall in the gap between the stalwart and the novice, as they are a group that can either become more committed or step back from action. The groups are important, because they either strengthen a movement by increasing the numbers who "bear witness" (Della Porta and Diani, 2006, p. 178) to increase "the number of participators in mobilizations which involve serious personal risk or cost" (2006, p. 178), or can spilt or break a movement as discussions in Chapters Four, Five and Six show that people often step back from direct action for fear of arrest, amongst other reasons.

Moreover, environmental activists have developed great skills in exploiting a gap between the climate skeptics, scientists and the moveable middle of the general public. This book takes the advertising motif of the "moveable middle" and applies it to understanding the strategies and tactics of ecoActivist movements. The moveable middle are people in society that are either at the fringes or not aligned to a cause and need to be informed to engage so a movement or collective can bear witness. Activists target the moveable middle to be able to build a groundswell of support from the bottom-up. A moveable middle example could be an individual against a single issue, or individual company. For example, an activist at the Camp for Climate Action (CCA) near Heathrow Airport, London in 2009 included local residents because, "they know at some level that big companies don't have their best interests at heart, so at some level they come on board with that and obviously, then, conversations happen" (Interview with HACAN Chair John Stewart, 2009). Researchers also note individuals are more motivated when there's a large-scale event. Saunders *et al.* (2012) note that when a protest group is deemed to have power, "protesters who identify with a group they consider more powerful are more likely to experience approach emotions, and consequently, are more likely to participate in protest" (2012, p. 7). With the Climate Camps for Action (CCA) although small scale at Drax (2006) the annual nature of the event meant by the time of the Heathrow CCA (2007), Heathrow the camp gathered a lot of momentum and support. Motivation for numerous reasons is one reason why people protest, but there are other factors. Chapters Five and Six explore this gap in more detail to show how some groups exploit the gap by firstly retaining an inside/outside dichotomy of social movements, secondly, what Saunders's *et al.* (2012) suggest is a gap in knowledge, so that "protest organizers might work harder to make calls to action more widely available through open channel" (2012, p. 19).

Kollmuss and Agyeman (2002) observe that individuals have both a value system and emotional connection with pro-environmental behaviour and they "see environmental knowledge, values, and attitudes, together with emotional involvement as making up a complex we call "pro-environmental consciousness" (2002, p. 256). Kollmuss and Agyeman (2002) also identify that people are motivated to protest when there are large scale actions, and if the action is something that fits within an individual's ideology and values. Kempton, Boster, and Hartley (1995) found that neither a prior knowledge of pro-environmental behaviour nor a comprehensive understanding of environmental science acts as motivators for pro-environmental behaviour.

Emotion plays an important, but not essential role in the motivation for people to undertake civil disobedience. Other factors are instrumental motives based on "the rational cost-benefits calculation of both collective incentives and selective incentives" (Saunders *et al.*, 2012, p. 8), where selective incentives are the material benefits for the individual, and whether any action will achieve a political change. Motivation also comes from a belief any action will lead to "some level of perceived political efficacy" (Saunders *et al.*, 2012, p. 8). Thus motivation and emotional strength combined with an understanding to capacity build the movable shows the highly organized nature of ecoActivists movement, regardless of where individuals are novices or stalwarts.

Alex Plows (2006) also notes the strength of "different networks and campaigns [that] share the same struggles" consolidated "the ground from which anti-globalisation action springs … through continued focus of generating and building capacity at local group level; eco camps are useful way of doing this" (2006, p. 26). Chapters Five and Six look in more detail on tactics and strategies taken by social movements to influence the moveable middle, by utilizing and upending technological and traditional media practices. Exploitation of this gap has enabled activist groups to circumnavigate and contest the mainstream media debates on the environment and retain a presence within the political discourse. Using citizen or alternative journalism, and the space of protest (see Chapter Seven) targets what's sometimes referred to by activists as moveable middle. Thus, whilst the political elite's solution to climate change is embedded in economics, ecoActivist garner support by targeting the moveable middle in their capacity building and value laden motivational actions. This war between economics and egalitarianism is played out in the traditional media platforms.

The Climate Science Behind the Headlines

UK researchers found that overall, news media coverage of climate change had raised awareness in public knowledge. Although, evidence for why climate change was happening, or how it would affect them was less clear. A Department of Environment, Food and Rural Affairs (DEFRA) report found that awareness of climate change had "been achieved primarily through a significant increase in the coverage of climate change by the news media" (Ward, 2009, p. 59). Environmental movements and scientists also had some contribution to the increasing awareness, mainly through "greater promotional and communication activities about climate change by some scientists, environmental groups, and policy makers" (DoE, 2008, p. 60). DEFRA's report concluded that "the public were becoming more aware of climate change, however many remain uncertain on the science, causes and modes of prevention" (2008, p. 60).

Boykoff and Boykoff (2007) and Boykoff and Mansfield (2008) found greater focus on mass media coverage of the climate change debate than the public attitude. A Cardiff University study (2015) conducted by IPSOS Mori on "Public perceptions of climate change in Britain following the winter 2013/2014 flooding"[5] found that overall an increase in concerns over climate change, increasing by 26%, up to 69%. Of those surveyed, there was a general belief that all forms of news media, (including TV, newspapers and the internet) were more influential at informing the public than scientific data (3% and less than 1% respectively). There was strong narrative that flooding and climate change were linked in some way. Fifty two percent of respondents agreed that "most media reports at the time linked the floods to climate change than did not" and 14 didn't believe there was narrative connecting the two events.).

In Australia, similar findings were made in a study of flooding in South east Queensland, Australia (2010/11). South east Queensland experienced its second worst floods in its history, when Tropical cyclone Yasi, a category five hit land in the northern and far northern regions of the state. The resulting deluge dumped vast amount of rain, which ran down into central and southern Queensland, flooding the state capital Brisbane. Accusations were made that management of dams were poor leading to a flash flood in Grantham across the southern half of the state claiming 33 lives. Bohensky and Leitch's (2014) systematic study on newspaper analysis of the floods aimed show a misrepresentation of the scientific process of measuring climate change led to a general lack of understanding of the facts. The confusion came in part, from "inaccurate sound bites which confound the media discourse on climate change science" (2014, p. 486).

In 2007 the media mogul, and News Corp International founder, Rupert Murdoch "publicly committed his media outlets to reporting climate change as a real threat and reporting it in new vivid ways. He also committed his corporation to become carbon neutral" (Cottle, 2009, p. 506). Murdoch is quoted as saying "Climate change poses clear, catastrophic threats. We may not agree on the extent, but we certainly can't afford the risk of inaction. The climate will not wait for us" (Osbourne, 2015). Yet despite what Cottle first observes that "climate change, it seems, has eventually come of age in the news media as a global crisis" (Cottle, 2009, p. 506), Murdoch's News Corp publications remain skeptical that climate change is a consequence of human industrialization. Quoted in his own newspaper *The Sun* newspaper, Murdoch largest newspaper in the UK quotes his as saying that

> we should approach climate change with great scepticism. Climate change has been going on as long as the planet is here. There will always be a little bit of it. We can't stop it. We've just got to stop building vast houses on seashores. The world has been changing for thousands and thousands of years, it's just a lot more complicated today because we are more advanced. (Cottle, 2009, p. 508)

This narrative is consistent in News Corps' Australian press. Media reporting of climate change science and environmental matters is influenced by the climate sceptic debates. Robert Manne's essay (2011) *Bad News: Murdoch's Australian and the Shaping of the Nation* found Australia's only general news national newspaper and owned by News Corp, *The Australian* coverage of climate change is "what *The Australian* has contributed on climate change under Chris Mitchell's [editor] watch is a truly frightful hotchpotch of ideological prejudice and intellectual muddle" (Manne, 2011, p. 40). McGaurr, Lester and Painter's (2013) study conclude that the climate change in the 2013 Australian Federal election coverage found that *The Australian* newspaper "is skewed by the fact that … [It] has maintained a particular position on climate change for a decade" (2013, p. 11). However, climate skepticism although prominent in News Corp newspapers, skepticism appears in other publications. Academic and journalists Wendy Bacon (2011) found that "coverage of climate change by ten Australian newspapers from February to July 2011 tended to be negative towards policy" (2011, p. 62). Bacon's follow-up study *Sceptical Climate part 2: Climate Science in Australian Newspapers* (2013) found that a third of the articles "did not accept the scientific consensus" (p. 45).

Overtime a small group of news-privileged climate change skeptic contested the anthropogenic claims about the environment (Boykoff and Boykoff, 2004, 2007; Monbiot, 2007). Climate deniers and sceptics have the attention of newspaper owners and as the studies show (Churchill, Eagle, Low, and Schurmann, 2015) that many climate sceptics and deniers will be given opportunities to write editorial pieces for favourable newspapers, be interviewed and present an argument based on emotive language. In addition, climate sceptic and deniers will use influential tactics of marketing, sponsorship and opinion pieces in national newspapers. A third way climate sceptics attempting to influence public opinion in by characterizing ecoActivists as either criminal, such as the neologisms ecoterrorism (Arnold, 1987) and/or Watermelon's—environmentally and politically green outwardly, and red (socialists) inside (Delingpole, 2011).

Climate skeptic Delingpole, who coined the term watermelon environmentalism in his book, *Watermelon; How Environmentalists are Killing the Planet, Destroying the Economy and Stealing your Children's Future*, views environmental activists as an enemy of big business and economic growth. Delingpole and other skeptics see all environmental groups from professional (Greenpeace, Friends of the Earth) to radical (Global justice movement) hold damaging socialist (watermelon) views. Klein (2014) in her interview with Delingpole notes that, "Modern environmentalism successfully advances many of the causes dear to the left: redistribution of wealth, higher taxes, greater government intervention, regulation." Klein adds that Heartland president Joseph Bast puts it even more bluntly. For the left, "Climate change is the perfect thing. … It's the reason why we should do everything [the left] wanted to do anyway" (2014, p. 42). Chapter Four takes a closer look at the various strategies that climate deniers and sceptics take in undermining and challenging the climate debate. Whether it is anti-environmentalists, media moguls, climate scientists, climate sceptics, climate deniers, journalist, the public or NGO, this book will show that over time the voice of the ecoActivists is missing from how the public receive and understand information about climate change and environmental issues.

Current Climate Debate, the Role of Traditional and Social Media

Today, in our technologically digital literate world machines allow us to support environmental journalism and political participation with the click of a mouse (clicktivism). Clicking social media likes, shares and virtual thumbs-up for some is activism. Others refer to this more as slacktivism a term coined by (Christensen, 2011), or clicktivism- clicking a mouse in support. The digitally literate can use social media to show support for a cause by changing their Facebook profile to the flag of a country, liking and sharing a post, or tweet and retweeting a charitable link or pledge for cash. These forms of clicktivism and slacktivism are passive "political activities that have no impact on real-life political outcomes, but only serve to increase the feel-good factor of the participants" (Morozov, 2009 cited in Christensen, 2011).

Social media provides, to some extent, a way of grassroots, radical and NGO environmental groups to have a voice and contribute knowledge to the climate debate. Chapters Three, Four and Five look at the environmental debates and discourse through the lens of social media. Social media inception from the mid noughties (2000) coincided with a greater coverage of environmental issues, and a rise in local and global justice and environmental protest movements. Facebook (ca. 2004) and Twitter (ca. 2006) opened up more channels of communication between activists groups, and shifting both the time and space of environmental protests. Chapter Four will explore how activists have engaged with social media to bypass traditional media practices, and circumnavigate gate keeping processes found in large multinational media organization.

Gerbaudo (2012) and Gitlin (2012) show how a technological change in media production and consumption sees a growth in social media usage alongside the retention of physical repertoires of protest (such as protest camps and their rituals). Social media provides a path to join the public debate, approaching the moveable middle, convincing the novices and returners to move towards becoming stalwart and repeaters. At global summit protests, social media has proved vital in coordinating and organizing a ground swell of bottom up support. However, the use of social media has become so prolific there's often too much focus on on-line presence over the physical act of protest. As one activists said at anti-fracking collective, Reclaim the Power's London gathering, "Once we have a website, we can start planning" (activists J, 2015).

What is emerging is a narrative that ecoActivism needs to be strategic to keep any position within media discourse. Power relations between business and science can be to the detriment of ecoActivists. With so many stakeholders, scientists, business, policy makers, environmental NGOs, social movements, climate justice movements and individuals, radical and non-hierarchical movements there are a lot of voices shouting about climate change and environmentalism. EcoActivists have found some ways to challenge the dominant positions, although social media and targeting the moveable middle, goes some way to having a voice, nevertheless there are many obstacles, from one-size-fits-all legislation, to economic arguments, negative media reporting, frames of violence and deviancy, undercover policing, surveillance tactics and data and metadata tracking. At each obstacle, ecoActivist have found innovative ways to ensure they have a voice in the mainstream media. As chapters five and six show, being creative can be very productive in generating media events.

Structure of the Book[6]

The first half of this book outlines the interconnectivity between media theory, green governmentality and political policy makers to show how the voice of the ecoActivists has been lost over time. Chapters' Two, Three and Four contend that in the battle for green capitalism, consumer-led environmentalism has become the acceptable position for environmental and political participation by supporting liberal democracies and neoliberals in applying economics to climate change solution. Furthermore, groups or individuals choosing to challenge the dominant position in favour of egalitarian approach through non-violent direct action or civil disobedience has given government justification to curtail and restrict activities because of national security fears.

The second half of the book examines how activists are reclaiming a media voice through tactical media strategies and explores the consequences for eco-Activists set out in the first half to examine how movements can influence public opinion in a mediatized world. Chapters Five and Six set out the ways in which environmental activists have contested media coverage and political policing to retain a voice in environmental discourse. Chapters Five and Six also explore these concepts and the application of social media as a strategy to shift power back towards a middle ground. However, although some academics and activists hail social media as helping to foster greater democracy, there's a risk of over-reliance on social media strategies, resulting in too much focus on hits and likes (see Chapter Six). Facebook and social media as a protest tool can equally be detrimental to wider democratic practices, and so Chapter Seven argues that the space of environmental activism is a much-more effective mode of contentious politics. Chapter Seven will explore the use of space and the space of protest camps to examine the fine line between alternative, counter society and the risk run by activists of mirroring political governance with an environmental lilt.

The book then moves onto looking at environmental activism from the local to the global stage. The book looks at the phenomenon of summit mobilization as large-scale protests centred around the Multilateral UN summits on climate change has become common place. The final chapter (Chapter Eight) looks at the future of the movement, in light of the Paris attacks and increasing security narrative around protest, where does protest go from here? Chapter Eight draws on interviews with those in Paris, and the UK, the Reclaim the Power and the move towards movement building plans for 2016 and beyond. This chapter draws together many of the strands in the book to explore the spectrum of environmental discourse from activism to political economy. The Conference of the Parties (CoP) of the United Nations Framework Convention on Climate Change draws many activists from around the globe along with the G20 and G8, meaning many have high security and new legislative measures that curtail protest.

CoP are opportunities for significant mega-protest (Death, 2010) for they provide platforms to explore media representation of environmental activism on a global scale. The following chapter takes a historical look at relations of power in media reporting, from Greenpeace (1971), to Greenham Common Peace Camp (1982–2000), to the anti-capitalist protest (1999–2010), and Camps for Climate Action (aka Climate Camps (2005–2010)) through the concept of media movements.

Notes

1. This book separates out the term media into three areas, traditional, social and activist. Traditional media refers to mainstream, established and traditional forms of media both online and offline, in print (newspapers); radio and television. Social media is built on a Web 2.0 platform as a "group of internet-based applications that build on the ideological and technological foundations" (Veil, Buchnart, and Palenchar, 2011, p. 505), such as Facebook, Twitter, Instagram and app-based sites. Activists' media ranges from independent news websites, such as Indymedia, VisionOn.TV, Climate camp media; Dissident island radio and listservs; but also includes flyers, leaflets and handbooks.
2. http://blogs.ei.columbia.edu/files/2009/10/broeckerglobalwarming75.pdf
3. http://www.legislation.gov.uk/ukpga/2011/13/section/143/enacted
4. Saffron *et al.* method meant data was collected over a fortnight for each working group (WG), centred on the day of the IPCC press release as follows: WGI: 21 September 2013 to 4 October 2014; WGII 24 March 2014 to 6 April 2014; WGIII 7 April 2014 to 20 April 2014. All media for analysis were selected on the basis of influence (circulation, audience) and ideology (2015, p. 384).
5. http://psych.cf.ac.uk/understandingrisk/reports/URG%2015-01%20WinterFlooding.pdf

References

Anderson, A. (1997). *Media, Culture and the Environment*, London: Routledge.

Anderson, J. (2004) "Spatial Politics in Practice: The Style and Substance of Environmental Direct Action", *Antipode*, 36(1), pp. 106–125.

Arnold, R. (1987). *Ecology Wars: Environmentalism as If People Mattered*, Bellevue, WA: Free Enterprise Press.

Bacon, W., Jegan, A., and Australian Centre for Independent Journalism (issuing body.) (2011). *Sceptical Climate*, Ultimo, NSW: Australian Centre for Independent Journalism.

Bacon, W. (2013). *Sceptical Climate Part 2: Climate Science in Australian Newspapers*, Ultimo, NSW: Australian Centre for Independent Journalism.

Bohensky, E. L., and Leitch, A. M. (2014). "Framing the Flood: A Media Analysis of Themes of Resilience in the 2011 Brisbane Flood", *Regional Environmental Change*, 14(2), pp. 475–488.

Boykoff, M. T. | Boykoff, J. M. (2007). Climate change and journalistic norms: A case-study of US mass-media coverage. *Geoforum: Journal of Physical, Human and Regional Geosciences.*, 38(6), 1190–1204.

Boykoff, M. T., & Mansfield, M. (2008). 'ye olde hot aire': Reporting on human contributions to climate change in the UK tabloid press. *Environmental Research Letters*, 3(2), 024002. doi:10.1088/1748-9326/3/2/024002

Broecker, W. S. (1975). "Climatic Change: Are We on the Brink of a Pronounced Global Warming?", *Science*, 189(4201), pp. 460–463. doi:10.1126/science.189.4201.460

Buckingham, S., and Turner, M. (2008). *Understanding Environmental Issues*, London: Sage.

Carson, R. (1962). *Silent Spring*, Wilmington, MA: Mariner Books.

Carvalho, A. (2007). "Ideological Cultures and Media Discourses on Scientific Knowledge", *Public Understanding of Science*, 16(2), pp. 223–243.

Chatterton, P., Featherstone, D., and Routledge, P. (2013). "Articulating Climate Justice in Copenhagen: Antagonism, the Commons, and Solidarity", *Antipode*, 45(3), pp. 602–620. doi:10.1111/j.14678330.2012.01025

Christensen, H. (2011). "Political Activities on the Internet: *Slacktivism* or Political Participation by Other Means?", *First Monday*, 16(2). doi:10.5210/fm.v16i2.3336

Churchill, M. W., Eagle, L. C., Low, D. R., and Schurmann, A. J. (2015). "Pro-environmental Behaviors of Climate Skeptics: Do Beliefs Matter? A Study of the Dry Tropics, Australia", Proceedings of the 7th International Conference on Climate Changes, Global Warming, Biological Problems (CGB'15). 5–7 September 2015, Seoul, Korea.

Cohen, I. (1996). *Green Fire*, Pymble, NSW: Angus & Robertson.

Cottle, S. (2009). *Global crisis reporting: Journalism in the global age / Simon Cottle*. (Issues in cultural and media studies). Maidenhead: McGraw-Hill/Open University Press.

Cottle, S., and Lester, L. (eds.). (2011). *Transnational Protests and the Media*, New York, NY: Peter Lang Publishers.

Couldry, N. (2000). *The Place of Media Power: Pilgrims and Witnesses of the Media Age*, London: Routledge.

Crutzen, P., and Stoermer, E. (2000). "The 'Anthropocene'", *Global Change Newsletter*, 41, pp. 17–18.

Death, C. (2010). *Governing Sustainable Development: Partnerships, Protests and Power at the World Summit.* Hoboken, NJ: Taylor & Francis

Della Porta, D., and Diani, M. (2006). *Social Movements: An Introduction* (2nd edn.), Mladen, MA: Blackwell.

Delingpole, J. (2012). *Watermelons: How Environmentalists are Killing the Planet, Destroying the Economy and Stealing Your Children's Future.* London: Biteback.

Deluca, K. (1999). *Image Politics: The New Rhetoric of Environmental Activism*, New York, NY: The Guilford Press.

Doyle, J. (2007). "Picturing the Clima(C)Tic: Greenpeace and the Representational Politics of Climate Change Communication", *Science as Culture*, 16(2), pp. 129–150.

Dryzek, J. (2000). *Deliberative Democracy and Beyond: Liberals, Critics, Contestations*, New York, NY: Oxford University Press.

Dryzek, J. (2005). *The politics of the earth: Environmental discourses / John S. Dryzek.* (2nd ed.). Oxford: Oxford University Press.

Dryzek, J., and Schlosberg, D. (2005). *Debating the Earth: The Environmental Politics Reader*, Oxford: Oxford University Press.

Fairclough, N. (1995). *Media Discourse*, London: Edward Arnold.

Fairclough, N. (2000). *New Labour, New Language?* London: Routledge.

Foreman, D. (2005). "Putting the Earth First", in J. Dryzek and D. Schlosberg (eds.), *Debating the Earth: The Environmental Politics Reader*, Oxford: Oxford University Press, pp. 348–354.

Foucault, M. (1970). *The Order of Things: An Archaeology of the Human Sciences*, New York, NY and London: Tavistock.

Foucault, M. (1972). *The Archaeology of Knowledge*, London: Tavistock Publishers.

Foucault, M. (1977). *Discipline and Punish: The Birth of the Prison*, London: Penguin.

Foucault, M. (1978). *The Will to Knowledge: The History of Sexuality:1*. Translated by R. Hurley, London: Penguin.

Foucault, M. (1982). "The Subject and the Power", in H. Dreyfus and P. Rabinow (eds.), *Michael Foucault: Beyond Structuralism and Hermeneutics*, Brighton: Harvester, pp. 208–226.

Foucault, M. (1986). *The Care of the Self: The History of Sexuality*. Vol. 3. Translated by Robert Hurley, New York, NY: Random House.

Foucault, M. (1989). "Of Other Spaces", *Diacritics*, 16(1) (Spring, 1986), pp. 22–27.

Foucault, M. (1991). "Governmentality", in Burchell *et al.* (eds.), *The Foucault Effect*, Hemel Hempstead: Harvester Wheatsheaf, pp. 87–104.

Foucault, M. (1993). "About the Beginning of the Hermeneutics of the Self (Transcription of two lectures in Darthmouth on Nov. 17 and 24, 1980)", in M. Blasius (ed.), *Political Theory*, 21(2), pp. 198–227.

Foucault, M. (1997). *The Politics of Truth*, Los Angeles, CA: Semiotext(e).

Foucault, M. (2002). *Power: Essential Works of Foucault*, London: Penguin.

Foucault, M. (2007). *Security, Territory, Population: Lecturers at the College De France, 1977–1978*, New York, NY: Picador, pp. 333–363.

Foucault, M. (2008).*The Birth of Biopolitics. Lectures at the Collège de France, 1978–79*. Edited by M. Senellart and Translated by G. Burchell, Basingstoke: Palgrave Macmillan.

Gamson, W., and Modigliani, A. (1989). "Media Discourse and Public Opinion on Nuclear Power: a Constructionist Approach", *American Journal of Sociology*, 95(1), pp. 1–37.

Gerbaudo, P. (2012). *Tweets and the Streets: Social Media and Contemporary Activism*, London: Pluto Press.

Gitlin, T. (1980). *The Whole World is Watching: Mass Media in the Making & Unmaking of the New Left*, Berkeley, CA: University of California Press.

—— (2012). *Occupy Nation: The Roots, The Spirits and the Promise of Occupy Wall Street*, New York, NY: HarperCollins.

Gore, A. (2007). *Earth in the Balance: Forging a New Common Purpose*, London: Earthscan Publications.

Hickman, L. (2015). "The IPCC in an Age of Social Media", *Nature Climate Change*, 5, pp. 284–286. doi:10.1038/nclimate2528

K. (2001). "Being Black Block", in Anon eds, *On Fire: The Battle of Genoa and the Anticapitalist Movement*, Edinburgh: AK Press.

Kempton, W., Boster, James S, & Hartley, Jennifer A. (1995). *Environmental values in American culture / Willett Kempton, James S. Boster and Jennifer A. Hartley*. Cambridge, Mass; London: MIT Press.

Klein, N. (2014). *This Changes Everything: Capitalism vs. the Climate*. London: Allen Lane.

Kollmuss, A., and Agyeman, J. (2002). "Mind the Gap: Why Do People Act Environmentally and What are the Barriers to Pro-environmental Behavior?", *Environmental Education Research*, 8(3), pp. 239–260. doi:10.1080/13504620220145401

Lester, L. (2010) *Media and Environment: Conflict, Politics and the News*, Cambridge: Polity.

Lester, L., & Cottle, S. (2009). Visualizing Climate Change: Television News and Ecological Citizenship. *International Journal Of Communication*, 3, 17. Retrieved from http://ijoc.org/index.php/ijoc/article/view/509

Luke, T. (1999). "Environmentality as Green Governmentality", in E. Darier (ed.), *Discourses of the Environment*, Oxford: Blackwell, pp. 121–151.

MacKay, G. (1998). *DIY Culture: Party and Protest in Nineties Britain*, London: Verso.

Manne, R. (2011). "Bad News: Murdoch's Australian and the Shaping of the Nation", *Quarterly Essay*, (43), pp. 1–119.

Martell, L (1994). *Ecology and Society: An Introduction*, Cambridge, UK: Polity.

Martin, G. (2015). *Understanding Social Movements*, London: Routledge.

McCormick, J. (1989). *The Global Environmental Movement: Reclaiming Paradise*, London: Belhaven Press.

McGaurr, L., Lester, L., and Painter, J. (2013). "Risk, Uncertainty and Opportunity in Climate Change Coverage: Australia Compared", *Australian Journalism Review*, 35(2), pp. 21–33.

Monbiot, G. (2007). *Heat How to Stop the Planet from Burning*. Cambridge, MA: South End Press.

Newell, P., and Paterson, M. (2010). *Climate Capitalism: Global Warming and the Transformation of the Global Economy*, Cambridge: Cambridge University Press.

Oels, A, Feindt, P. H, Oels, A. (2005). Rendering Climate Change Governable: From Bio-power to Advanced Liberal Government? *Journal of Environmental Policy & Planning.*, 7(3), 185–208.

Osbourne, H. (2015). Rupert Murdoch buys National Geographic: Quotes from media mogul on climate change. Available at http://www.ibtimes.co.uk/rupert-murdoch-buys-national-geographic-quotes-media-mogul-climate-change-1519188

Perez-Pena, R. (1992). "Gladwin Hill Dies at 78; Times Reporter 44 Years". Available at http://www.nytimes.com/1992/09/20/us/gladwin-hill-dies-at-78-times-reporter-44-years.html

Plows, A. (2006). "Blackwood Roads Protest 2004: An Emerging (Re)Cycle of UK Eco-Action?", *Environmental Politics*, 15(3), pp. 474–484.

Saffron, Hywell, Kurz, Wiersma, and Boykoff. (2015). "Dominant Frames in Legacy and Social Media Coverage of the IPCC Fifth Assessment Report", *Nature Climate Change*, 5, pp. 380–385. doi:10.1038/nclimate2535

Saunders, C., Grasso, M., Olcese, C., Rainsford, E., and Rootes, C. (2012). "Explaining Differential Protest Participation: Novices, Returners, Repeaters, and Stalwarts", *Mobilization*, 17(3), pp. 263–280.

Smith, R. (1872). *Air and Rain. The beginnings of a chemical climatology*. London

Thomas F. Stocker [and nine others] (ed.). (2014). *Climate Change 2013: The Physical Science Basis: Working Group I Contribution to the Fifth Assessment Report of the Intergovernmental Panel on Climate Change*. New York, NY: Cambridge University Press.

Torgerson, D. (1999). *The Promise of Green Politics: Environmentalism and the Public Sphere*. Durham, NC: Duke University Press.

Van Dijk, T. (1988). *News Analysis: Case Studies of International and National News in the Press*. Hillsdale, NJ: Laurence Erlbaum.

Veil, S. R.,Buehner, T.,Palenchar, M. J. (2011). A Work-In-Process Literature Review: Incorporating Social Media in Risk and Crisis Communication. *Journal of Contingencies and Crisis Management.*, *19*(2), 110–122

Viejo, E. (2003). "Protecting the Movement and Its Unity: A Realistic Approach", in Notes from Nowhere (ed.), *We Are Everywhere: The Irresistible Rise of Global Anticapitalism*, London: Verso, pp. 371–374.

Ward, R.(2009). Climate Change, the Public, and the Media in the UK: A Watershed Moment, in Boyce, T., & Lewis, Justin. (2009). *Climate change and the media / edited by Tammy Boyce & Justin Lewis*. (Global crises and the media; v. 5). New York; Oxford: Peter Lang. pp. 59–65

· 2 ·

MEDIA GOVERNANCE, ECOACTIVISM AND THE TRADITIONAL MEDIA

Over the decades, ecoActivists groups and many other protest collectives have developed innovative ways to engage and even disengage with the traditional media. From the Greenham Common Women's Peace Camp (1982–2000), Greenpeace media strategy with the Royal Shell Oil Brent Spa rig (1995), Anti-roads protest (1990s) including the trope of Swampy (1996/1997), Global Justice Movement/Anti-capitalists protest and May Day Protests (1999–2004, 2009/2010), and Camps for Climate Action (2005–2010) all had different ways of engaging with the traditional media.

The final decade of the twentieth century (1990s) was a tipping point in the relationship between ecoActivism, traditional media corporations, and politicians. Since the nineteen seventies, public and political attitudes on environmental issues were in the main, shaped by Environmental Non-Governmental Organizations (ENGO) and radical activism. However, ENGO's and radical voices were increasingly facing discursive challenges from politicians, international organizations, and multinational corporations beginning to embrace green markets in seeking solutions to climate change. Deregulation of media markets meant that neoliberal and conservative political rhetoric had a new platform. New satellite and cable broadcast channels predominantly owned by News Corp. Sky News UK and Fox News (USA), Star News (India) provided a platform for the climate skeptic right.

Consequently, journalists began turning away from ENGO and radical activists for climate change information and towards politicians and big business as opinion leaders on climate issues. This slow move away from ENGO and radical activists was, as this chapter will argue, part of the green governmentality that saw a shift in media voices on climate change ultimately resulting in the quietening of the ecoActivist voice in the UK press.

Survey data from Europe and the USA found a significant increase in environmental concerns between the mid-1960s and early 1970s, spiking again in the 1980s. In 1989, UK data reveals that 35 per cent of "the public saw the environment as an important issues facing the country" (Anderson, 1997, p. 91). Environmental non-governmental organizations (ENGO) were also growing during this period. ENGO played a key role gathering environmental data, and used as "evidence about key environmental issues later acknowledged by governments and official scientific institutions" (Grove and White cf. Anderson, 1997, p. 100). One of the world's first ENGO Greenpeace, had grown globally, and opened their international headquarters in the Dutch capital, Amsterdam (1989). Within six years, Greenpeace International had 2.5 million supporters in 158 countries. Their total net global income was $152,805,000 (USD) (1995) per annum.

Despite the growth of ENGO, there remained a perceived lack of action for some over degradation of the environment. The opening up of trade routes across Europe and the USA helped give momentum to the anti-road/anti-capitalist movements, as activists pushed back against the environmental damage being done by expanding multinationals. Under new transnational trade deals. The result is a split between the media and politicians advocating for economic growth and call for a more socially inclusive approach of environmental activists. A backlash begins to grow against economic and ENGO dominance of the environmental debate. Protests began moving away from the anti-nuclear protest camps of Greenham Common, Aldermaston (UK), Pine Gap (Australia), and Alaska (USA), towards similar action seen at Lake Pedder and Franklyn Dam protest in Tasmania, Australia (1970) and Earth First! in the USA (1979). Earth First! UK (1991) later becomes part of the anti-roads environmental and class war movements. Civil disobedience, non-violent direct action, civil unrest, protest, and ecoActivism spiked for the first time on UK news agendas between 1991 and 1997. Prior to this period, the largest site of contestation was the Greenham Common Peace Camp (1982–2000). Protest camps over road and airport expansion grew across southern England and throughout the UK. Treehouses and tunnels became the favourite tool of the anti-roads protests movements. The trope of

Swampy was born in Devon (1996–1997), and repeated at Manchester Airport over plans for a second runway (1997). In cities across the UK, Reclaim the Streets' free music parties and protest (1995–1996) would converge to reclaim the city space for parties, protest, and people.

This chapter charts the ecoActivists voice throughout this period to understand the position and relations of power between journalists and activists today. Unlike ENGO's such as Greenpeace and Friends of the Earth, and Sea Shepherd Conservation Society, the new social movements in ecoActivists lack resources mobilization (RM) of larger NGOs and rejects political lobbying in preference for a "perspective [which] assumes that a requirement for political influence is a change in organisational form that takes social movement organisations nearer the stereotypical interest groups. This approach—mainly in the RM mode—sees an almost inevitable process of development in which amorphous and fluid social movements become centralised and formalised" (Jordan and Maloney, 1997, p. 56), such as how the tactic of civil disobedience through direct action enabled ENGO and ecoActivists to be a leading voice (primary definer in Hall's terms, 1978) on environmentalism issues in the traditional media

Primary definers are journalistic sources given the role of a leading authoritative voice. Primary definers are often representations of authority, from the emergency services, politicians, experts, and commanding voices. The primary definer often also acts as a conduit to "reproduce symbolically the existing structure of power in society's institutional order" (Hall, 1978, p. 58). In general, a quote from a primary definer three or fourth paragraphs into the story, journalistically gives weight and authority to a news story. Hall argues that the relationship between the media and "primary institutional definers" gives the primary definers a chance to define an "initial definition or primary interpretation" of a particular topic. Therefore when the primary definers set the definition, they also "set the terms of reference within which all further coverage or debate takes place" (Hall, 1978, p. 58). Conflict in a story comes from the positioning of primary definers, often "setting up of a topic in terms of debate within which there are oppositions and conflicts is also one way of dramatizing an event so as to enhance its newsworthiness" (Hall, 1978, p. 58). In cases of a single primary definer, the story becomes unbalanced with the journalists favouring one side over another. Thus this chapter explores where the activists voice sits within the primary definer space of a news story. Identifying the primary definer enables us to identify the level of authority and credibility journalists and editors give to the activists voice. However, before we look at the activists voice through a series of case studies, the chapter looks more broadly at the field and the concept of primary definer.

Media, Culture, and the Environment—Revisited

Academic literature specific to environmental protest and media representation has steadily increased since Alison Anderson's (1997) seminal work, *Media, Culture, and the Environment*. Anderson's work is the first book to examine the relationship between environmentalism protesters and the traditional media. Anderson's (1997) Lester's (2007, 2010), Doyle (2007, 2009), Boyce and Lewis (2009), Painter (2013), Cottle and Lester (2011) and Lester and Hutchins (2013) each reflecting and examining the increasing interest in the environment, protest and the media. More general key texts on climate change and the media come from Boykoff and Boykoff (2007), Boykoff and Mansfield (2008), Doyle (2007, 2009,20166); Painter (2013); Boyce and Lewis (2009).Whilst works by Gitlin (2012), Klein (2014), Gerbaudo (2012); Lester (2010, Lester and Hutchins (2013), Frenzel, Feigenbaum, and McCurdy (2014), and Saunders (2013) focus on radical climate activism. Alison Anderson's significant work to the field, *Media, Culture, and the Environment* (1997) looks at how the media socially construct the environment Anderson's work offers an insight into the working practices of journalists, NGOs, and social movement organisations to show that environmentalism in the 1990s, although important to news agendas, often struggled to retain a place. Anderson's analyse examines the ways discourse shapes news by relying on taken-for-granted assumptions in news practice. Through example of the Greenpeace Brent Spa episode and a national newspapers campaign to save seals from a plague she unpacks questions of ideology, objectivity and impartiality to ask if media practitioners can be impartial or independent when targeted by environmental pressure groups. Anderson's empirical research consists of face-to-face interviews with leading environmental journalists to reveal the key characteristics of environmental news items. Whereas, this book takes a similar historical look at environmental discourse, it does so through the ecoActivist lens.

Boyce and Lewis's (2009) edited collection of media coverage on climate change essays, focusing on news coverage, creative fiction, and audience participation, alongside the growth of online media practices. The influence of politics and policy makers in both the UK and Australia shows how close media coverage and politics convene to frame climate change issues. Underpinning the essays is how the world's media frames complex climate changes issues impacts on public understanding as climate change became more prominent in the media by the start of the 21st century. These key text show how media governance is heavily influenced by government lobbying. Lobbying shapes a public discourse that understands climate change solutions as a political not science issue. Environmentalists rely on the science for their arguments and credibility (Anderson, 1997); but it is a political discourse, not scientific that has always and continues to frame the environmental debate in the media. Much of the work on climate communication focuses on science as discussed in the introduction. The idea of being wedded to climate science remains the backbone of the current climate change debates. Anderson (1997, p. 211) observations that environmental movements increasingly putting less stock into the science holds true today, leaving much of the climate science debate to policy makers and political economy arguments. With sceptics arguing against the science, and advocates both primary definer institutions (UNFCCC, IPCC; Government) are all using science as the foundation of their argument. Whereas, much as the ecoActivist's argument is embedded in political discourse and political economy. A brief scan of slogans at any climate change protest, are calling for political change, with the most common being 'System Change not Climate Change'; 'Green Jobs', or 'Climate Vote to Save it'.

Lester's work shows the dawning of environmentalism and protest in the news media (1980s) reveals an "increasing coverage of environmental issues in their bulletins and news pages, and establishing environmental rounds" (Lester, 2007, p. 1). Driving the coverage was a combination of political, cultural, and social discourses. Lester's follow-up book, *Media and Environment: Conflict, Politics and the News* (2010) argues that as a former environmental journalist she found it easier to get the word spondulicks into a news story than an environmental story into the news agenda. Like Anderson, Lester shows us how much news media frames shape public understanding of environmental issues. In examining the political and cultural discourse in environmental reporting, Lester finds environmental stories move lower down and even off news agendas on a regular basis. This side stepping of environmental stories or giving them less prominence is according to Lester a tactic to favour less contentious stories. Consequently, public understanding of environmental conflict in the media and it's relation to the everyday then draws out questions of "our shared understandings of place and community, or local responsibility and global citizenship, and about how individuals and societies act in ways that affect how we- and others- live now and in the future" (Lester, 2010, p. 7).

Hansen's *The Mass Media and Environmental Issues* (1993) and *Environment, Media and Communication* (2010) looks at how media messages come to shape our understanding of the environment and the defining of environmental problems. Hansen (1993, 2010) draws upon the experience of different countries to reveal how different theoretical approaches can unravel the relationships between the meaning of the environment and the media. James Painter (2013) looks at the representation of risk in reporting climate change found that much of the media coverage falls in to a discourse of disaster. In examining risk and risk relations to climate change. Painter's *Climate Change in the Media: Reporting Risk and Uncertainty* found that in media coverage of IPCC reports found that a disaster/implicit discourse "stresses the negative impacts of climate change and in its extreme form is seen as an obstacle to pubic engagement with the issue … [and] a barrier to public understanding and engagement" (2013, p. 136). The emphasis on negative aspects of the climate debate came partly from the scientists themselves, and partially from journalists negative reporting scientific data.

Authoritative Voices, Gatekeeping, and Activists as Primary Definers

Who speaks for whom and who journalist's source information from defines how the media influences the public. The source chosen by the journalist frames the story, making the language objective and fitting with the political or ideological position of the publication. This one-to-many media production model sees is a top-down flow of information that passes through gatekeeping processes. Gatekeeping is the process information moves through a series of gates before becoming public knowledge. Prior to an event any news value is given and changed as "a story is transmitted from one gatekeeper after another in the chain of communications", and managing the gates are "gatekeepers of information" (Dexter and White, 1964, p. 163). Gatekeepers can be journalists, editors, sub-editors, media moguls, public relations companies, advertisers, politicians, spin-doctors, councillors, community services, and all forms of broadcasting companies. Gatekeepers use a language that gives one objective view and become what Stuart Hall defines as terms Primary and Secondary definers (1973).

Primary definers are individuals associated with power and control. In selecting a source, journalists and editors decides if a primary definer is the one holding power, and therefore holds control of the information to shape the story. In daily news cycles, primary definers might be a police officer, politician, government spokesperson, -anyone the journalists or editor sees as an authoritative voice. A secondary source is someone of relevance to a story, but with less power and control. Taking Hall's ideas of primary and secondary definers we can see that primary definers are the key authoritative voice and given priority in the authentication of a story. In placing a person of power as an authoritative figure at the top of the story gives that person and their comments more authority, which then shapes public understanding (known as discourse). Yet discourse "never consists of one statement, one text, one action or one source" (Hall, 1973, p. 293), as it is part of a set of statements that is "characteristic of the way of thinking or the state of understanding at any time" (Hall, 1973, p 293). Add together a primary definer, with gatekeeping practices, you get to what Becker terms a "Hierarchy of credibility", which is "in any system of ranked groups, participants take it as given that members of the highest group have the right to define the way things really are" (1967, p. 241). Hence, when a protester is quoted ahead of say a representative of the states, then the protester controls the hierarchy of credibility, with the journalists seeing the protester as the most authoritative voice. Once a source is clearly defined, the hierarchy of credibility shapes the authoritative voice, identifies key mobilizations, and the actor sets the frame.

Secondly, understanding a story of an event, group, network, or action depends on the structural nature of the text. News language and meaning around protests is important in understanding representation in the present. Halloran, Elliott, and Murdock (1970) analyzed news coverage of the anti-Vietnam protest (1968) drew on Lang and Lang's (1955) study of four television news outlets covering a local election in American. Lang and Lang, found reporting often has an "inferential structure", a preferred meaning, a set pattern, a routine way of reporting. Lang and Lang found that both intentional and unintentional bias by the reporting "can influence public definitions in a particular direction" (1955, p. 171). Halloran *et al.* (1970) using content analysis to find out if newspapers used an inferential structure to test if protesters were seen in a favourable, unfavourable, or neutral light.[1] Marchers were represented as young, violent anti-American hooligans giving "an indication of the way in which ideas about current events are structured, simplified and fed into the general social consciousness" (Halloran *et al.*, 1970, p. 216). Specific descriptions included the terms "militant", "activist", "student", and "left-wing". Unfavourable terms included "confrontation", "attack", "riot", "mob" and "extremist" (1970, p. 108). Equally, Halloran *et al.* (1970) found less aggressive adjectives, such as "peaceful", "sincere" and "good humoured". As we will see militant riot mob and extremism are all adjectives repeatedly used to frame environmental protest.

An ability to understand the framing of an article equips the reader with the means to identify the "central organizing principle that holds together and gives coherence and meaning to a diverse array of symbols" (Gamson, Croteau, Hoynes, and Sasson, 1992, p. 384). Carvalho suggests such framing within "media packages work[s] as an overall interpretative principle in relation to the issue" (2000, p. 9). Subsequently media frames represent cultural codes or maps of meaning that places an event within a set of values (Hall, 1997). Maps of meanings are ways meaning is gained from pre-supposed, pre-shaped cultural knowledge (Hall, 1997). Cultural codes shape our knowledge, which we express through language and other forms such as hand gestures, body language, and stereotyping. How and why journalists apply a frame to a news story either reaffirms or challenges a set of cultural codes in news production. Gamson and Modigliani (1989) show how we interpret codes, through frames, each frame are a series of codes, which direct and guide the reader/viewer to interpret the information. Ways of framing are metaphors, exemplars (i.e. historical examples from which lessons are drawn), catchphrases, depictions and visual images (and semiotic analysis), visual images, photographs, and video acts all work as reasoning devices (and thus justification) within cultural codes to construct a story.

Carvalho's (2000) study of news story mechanisms are categorized into six areas of analysis (1) surface depicters and structural organizations, (2) objects, (3) actors, (4) language and rhetoric, (5) discursive strategies and processes, and (6) ideological standpoints (2000, p. 22). In analysing media reporting of climate change at the United Nations Conference of Parties (CoP), Carvalho identifies objects in a text being the theme of "economics, government, or nature, while more specific ones [topics] may be, for instance, climate change impacts on agriculture". By identifying 'who' an article mentions (the actors), the recognition of any Carvalho notes "individuals and institutions. Social agents and characters" provides the tools to unpack the role and the "perceived influence in shaping the overall meaning of the text are all elements of discourses within a text" (2000, p. 24). The actors then become the primary definer, "by having them represented by journalists either in the form of quotes or regular text" (Gamson *et al.*, 1992, p. 384). The construction of a story for public consumption is significant to understand the "cognitive process involved in news production and decoding" (Carvalho, 2000, p. 6). Cognitive processes help us decipher who is an important element and who is not an authoritative figure in a story. In other words, what is sayable and unsayable, who speaks for whom, and which voices (actors) frame the story (the discourse), primary definers, and ultimately public understanding.

A discourse, put simply, is the way that language gives us meaning about a particular historical moment. To understand how and why language provides a set of knowledge we need to look at the "production of understanding through language" (Hall, 1973, p. 291). How a word gains meaning, and the timeframe in which it exists, provides a position that the public take as understood to be a true account of a historical moment. Once we have a true account, we then base all future references on that discourse. Language shapes meaning by providing tools that help us comprehend the "forms the locus of tradition, of the unspoken habits of thought, of what lies hidden in people's mind" (Foucault, 1970, p. 297). Techniques such as reinforcing a frame through discourse "accumulates an ineluctable memory" (Foucault, 1970, p. 297) starts from a generally agreed upon discourse from language a horizontal/transparent position (Foucault, 1970). Through the concepts of primary institutional definers, and set terms of reference we move to a vertical/opaque position, where meaning is shaded and unclear, and subjective by the primary definers that gives us a discourse.

Language works as building blocks to structure meaning for us to be able to understand the construction and reasoning of any topic. Eventually, through different discursive challenges a word, as language takes up a position to become an acceptable meaning and general understanding. The result means that a meaning is given when all the building blocks are assembled. However, any meaning "depends on the relationship between things in the world" (Hall, 1997, p. 18), so that "language provides the tools for discourse as a system of representation" (Hall, 1997, p. 44). For a journalist the language they choose (positive/negative, active/passive) and the primary definer will frame a story in a certain way that both reinforces our pre-existing beliefs and gives us the tools to reading and interpretation of an event, person, idea etc. The language of a news story shapes how the public understands an event, issue, or process. Thus, as the rest of this chapter will show, by charting the language of the protester's voice in the UK newspapers we can see it begins with a strong voice from the ecoActivist dwindling over time as mainstream politics begin engaging in environmental rhetoric.

Media Language, Method, and the Politics of Protest

Understanding why and how activists are recast as a threat, (c.f Luke, 1999) this book's analysis is based on a combination of media theory and Critical Discourse Analysis (CDA) (Fairclough, 1995). Newspaper analysis over the four decades provides a way to chart the trajectory of the activists' voice. Newspapers were the most consistent form of media outlets, and therefore chosen as the main source of analysis. Fairclough's CDA differs from other kinds of discourse analysis is in his idea of "intertextual analysis" that works as a bridging tool between text and discursive practice. Intertextual analysis provides the tools to "look at the text from the perspective of discursive practice" (Fairclough, 1995, p. 70), as opposed to media practice (cf. Gamson *et al.* (1992), and Carvalho (2000)). Fairclough provides nine terms as a framework for reading the text in CDA. The categories are heterogeneity, grammar, boundaries, metaphor, nominalization, genre, glocalization, order of discourse and image. The heterogeneity element of CDA (Fairclough, 1995, p. 15) reveals activists are quoted through a mixture of linguistic styles that are predominantly informal, thorough colloquial language. Whereas, official sources are quoted in more formal, official language. Nominalization is the use of noun-like terms to identify those involved (such as women, protesters, anarchists), while action refers to the use of adjectives to define the events; and image is also considered important for Fairclough.

Fairclough links the use of metaphors to his idea of intertextuality, along with transformation. Agency is identified through the order of discourse (such as political, economic, environmental, radical politics); glocalization from the local to global—in which context is discourse placed; Boundaries concerns who is represented, who are journalists representing and which discourse is dominant. The last term is grammar, specifically the use of conjunctive terms. These last four; agency, glocalization, boundaries, and grammar are important and will now be looked at in detail.

CDA helps see how the media categorize one section of society, i.e. youth culture, into a genre. Other important variables in a text include the presence or "absences" and boundaries. Boundaries are often defined through a text as being between those who are being represented, and those doing the representing, when "boundaries are maintained between the representing discourse and the represented discourse between the voices of the reporter and the person reported" (Fairclough, 1995, p. 81). Boundaries also aid the analysis of absences in a text to discover "things which might have been 'there' but 'aren't" (Fairclough, 1995, p. 106). Fairclough terms "nominalizations" as the "processes that have been turned into noun-like terms (nominals) which can themselves function as participants" (Fairclough, 1995, p. 112). Nominalization is a type of "grammatical metaphor" (Fairclough, 1995, p. 220). This is similar to Hall's signification or "amplification spiral" that "suggests an increasing of deviances" (Hall, 1973, p. 223) by labelling individuals under the generic noun of "activists", as opposed to distinguishing which collective or organization the "activist" is connected too.

Genre can be "described in terms of organizational properties" (Fairclough, 1995, p. 56) that are defined through analysis of the "language associated with and constituting parts of some particular practice" (Fairclough, 1995, p. 56). The juxtaposing of one genre against another, results in several discourses forming into a single category. In very simplistic terms, this could be the genre of youth culture juxtaposed with crime discourse, equalling a category of youth crime. What is central to Fairclough's approach is the deconstruction of language to unpack how a journalist's use of language to create genres and categories (media discourse) affects the wider socio-cultural (media) practice.

An "order of discourse", like the Foucauldian order of discourse, is a means of highlighting "the relationship between different types in such sets", for instance, in a school, the discursive types of the classroom and the playground (Fairclough, 1995, p. 55). By drawing on specific communicative events over the forty-year period of environmental activism in the UK, this chapter will illustrate how historically orders of discourse have been "routinely used within the media" and how this "plays a part in the reproduction of the media system" (Fairclough, 1995, p. 72), by analysing power relations between the state and environmental activists. By developing Fairclough's notion of order of discourse with other CDA mechanisms, will, unlike Carvalho's study, (a) reveal which actors/voices are heard, and (b) show the power relations between actors. Other functions of the "order of discourse" is identification of "specific discursive practices of the text" (Philips and Jørgensen, 2004, p. 72). Analysis combining a communicative event with the order of discourse creates a dialectical position between language and social systems, with the journalist facilitating the conversation between activists, the state, and public. Philips and Jørgensen (2004) note that when a journalist draws on language "routinely used within the media, he or she also plays a part in the reproduction of the media system" (2004, p. 72). Reproducing traditional media messages is further enhanced through "mediated quasi-interaction" (2004, p. 72). The "order of discourse" allows for "one domain of potential cultural hegemony, with dominant groups struggling to assert and maintain particular structure within and between them" (Fairclough, 1995 cited in Philips and Jørgensen, 2004, p. 56). Therefore identifying the order of discourse in a communicative event over a historical period, and taking on board the journalist quasi-interaction, is useful in unpacking media discourse. Thus, in undertaking a historical analysis of ecoActivism reporting from the 1980 to the 2000, the study will illustrate that as environmental governance increases in social and political discourse. Only by charting key environmentalism-centred protests over a long period we can identify if any shift in power relations that affect ecoActivist role in shaping public opinion on climate change.

Combining media analysis with Fairclough's Foucauldian inspired critical discourse analysis enables us to see the how (media analysis) and why (critical discourse analysis) journalistic, political and activists language shapes public understanding of a subject. Drawing on media and cultural theorists (Fairclough, 1995; Carvalho 2000; Gamson *et al.* 1992; Gamson and Modigliani, 1989) the analysis identifies the positioning of the activist within the story as a primary definer (Hall, 1978, p. 58).

The findings show that early on, ecoActivist movements set the news agenda. Women at the Greenham Common Peace Camps (1980) were primary definers.[2] The language of Swampy and friends as primary definers defined the 1990s anti-roads movement. Swampy was also associated with a weekly newspaper column in the *Daily Mirror* newspaper (August 1997). There were Sunday supplement features on various activists, and there was even an ecoActivists character in a popular soap opera Coronation Street. A character Geoffrey 'Spider'. Nugent was a parody of Swampy, first appearing on December 15th, 1997, and recently returning to the soap opera as an off screen character. However, as this and the following chapter shows, post 1997 news agendas begin replacing the activists as "primary institutional definers" (Hall, 1978, p. 58) with police officers, and politicians and local councillors. At the same time all terms of reference of future coverage begin to change towards a certain discourse that places activism within a signification spiral amplified as deviant.

Thus, an activists' voice becomes quelled as a discursive challenge over the meaning of environmentalism, sustainability and green politics EcoActivism is framed as a negative engagement of environmental discourse, contrasted against at the time, was a need for more economic led solutions established in the non-binding the Kyoto Protocol (1996/1997) and increasing political acceptance of global warming and climate change (1997).

The next section will look at newspaper reports of three historical critically discursive moments of ecoActivism. A critical discursive moment is one that Carvalho defines as events or moments which involve "specific happenings that can lead to challenges to the 'established' discursive position" (2000, p. 5). The following pages will examine a series of case studies of what this book is defining as critical discursive moments within the history of ecoActivism. The critical discursive moments are The Greenham Common Women's Peace camp (1982–2000), the trope of Swampy and the anti-roads movement (1996–1998) the Global Justice Movement with a focus on the May Day (1999–2005) and G20 (2009) protests as key moment that shift the discourse and relations of power between ecoActivists, traditional journalism practices and the politics of protest.

Protest, Primary Definers and Discourse: An Example of the Reporting on Greenham Common Peace Camp

The Greenham Common Peace Camp was a protest camp against the storing of American nuclear warheads at the Royal Air Force (RAF) Base Greenham, Berkshire in south east England. It was embroiled in Cold War and hostilities between the USA and USSR. The UK/US military alliance led the UK to provide RAF Greeham as a base to house America's nuclear weaponry within Europe. The peace camp was a series of smaller camps outside each of the gates to the RAF base. Newspaper reporting and journalistic language framing the Greenham Common Peace Camp mobilizations often place the women mobilizations in a discourse of violence and aggression. News reports on confrontation uses words such as "force", "blockade", "tear down" and "bare hands" (Eldridge, 1995, p. 327). With much of the "violence against the women" going unreported and any police violence did "not make the news" (Eldridge, 1995, p. 327). Only when a group of female protesters trying to stop a delivery lorry from entering the base, did the media report on the action with a focus on "the police view: the men and supplies" (Eldridge, 1995, p. 327). Gaining media coverage of the protest had been problematic since the beginning of the Greenham Common campaign.

The peace camp began with a march (26 August 1981) from the Welsh capital Cardiff to Greenham Common, Berkshire, which had been a base since 1942. The hundred-mile march (160 km) by 40 women, some pushing prams others carrying anti-nuclear banners and flyers, was largely ignored by the traditional media. Eldridge (1995) found the first major traditional media report was almost 18 months after the camp began. The first large-scale action, Embrace the Base action, saw the women form a human chain outside RAF Greenham Common and the women, "were seen on television embracing the base" (1995, p. 318). Newspapers carried the image on their front pages as protesters held hands around the perimeter fence. The most newspaper coverage came from the mid-range papers, *Daily Express* and *Daily Mirror. The Daily Mirror*[3] headlines read "Peace: the plea by 30,000 women" (Palmer, 1982); "Peace war that won't go away" (Pratt, 1982); and "Scuffles as women seal off airbase" (Brown, 1982). The *Daily Express* framing linked the action to the cultural codes and maps of meaning associated with the cold war. The *Daily Mirror* front page is the visual imagery of an older woman with the peace sign painted on their faces. The front page images are accompanied by smaller images of women and children holding hands to embracing the base has the sub-headline, "Keeping warm" (Palmer, 1982, p. 2). Framing also evokes nationalism as "amid the sleet and rain of a bitter winter afternoon, 30,000 staged a remarkable protest [and] they linked hands in the heart of the English countryside" (Palmer, 1982, p. 1), compared to the "90 American cruise nuclear missiles" (Palmer, 1982, p. 1). Metaphors of war and peace see "a mass shout of freedom went out to signal the world's most powerful demonstration of peace" (Palmer, 1982, p. 11).

The journalists' language combines colloquialism with some aggression in the use of war metaphors, to define the peace camp. The order of discourse places the women and visitors as primary definers. Although not directly quoted, their descriptions frame the story to favour the activists. An absence in representation from the state, police, military police, or government reinforces the women as primary definers. One protester is described as a "mother of two Geraldine Adams from north London hung up nappies said they symbolize women in home" (Palmer, 1982, p. 1). Activists are personalized and named directly as "Rebecca Johnson, 28, who has camped outside the base for five months" (ibid). There is a reference to family unity as "seven year old Georgia Brown who was there from Oxford with her dad and two sisters", and "Pensioners Pat and Fred Sweeny who had travelled from Bognor Regis". Pat and Fred were joined by "social worker, Jean Wilson" and "Mary a militant multiple sclerosis sufferer" (Palmer, 1982, p. 3). Each of these identities (or actors in sociological terms) works as a bridge between the text and a discourse of unity (intertextuality in Fairclough's terms). The textual analysis of the Embrace the Base coverage shows support by the media for the peace camp through the different acknowledgment of actors and the descriptions of their everyday activities. However, this is not replicated in the second action, Dancing on the Silo's (New Year's Eve, 1982)

We find that media coverage of the action aligns with Eldridge's findings (1995). The Dancing on the Silos (01/01/1983) was an action where the women broke into RAF Greenham and danced at dawn on top of the nuclear weapon silos. Reports of the silo dancing are initially similar to the "Embrace the Base" coverage. The women remain primary definers, with the first quote from "Nell Logan, a 73 year old grandmother, whispered to me that she was a little nervous, but determined to go over the top" (Smith, 1983, p. 4). At paragraph ten (of 14), there is a second activist quote "they want to get them [the silos] finished before there is an election where there is a prospect of the missiles being rejected, said Deborah Law" (Bishop, 1983, p. 2). However, these two quotes are bookended with darker comments and almost a sense of subversive activity by the ecoActivists- which may be true.

The *Daily Express* newspaper uses language that describes the women as "dark figures of women in anoraks, trousers, and scarf's emerged from the Berkshire woods where they were sleeping" (Smith, 1983, p. 4). The underbelly language of "dark figures" could be seen as subversive action, but equally it can also be read as mythical metaphor of fairies and goblins sleeping in the woods. Describing "women in anoraks, trousers, and scarf's" is hardly the balaclavas of Black Bloc. Given the context of the peace camps' early days' of the Cold War, unsurprisingly a war metaphor remains throughout the reporting. Yet, now the women are the antagonists, findings that reflect Eldridge's study. The language reinforces the warlike action as "silently, they pulled hidden metal ladders from under the gorse bushes and advanced against the target" (Smith, 1983). Words such as "targets", "over the top" and "bring down", yet all localized in the "Berkshire woods" (Smith, 1983, p. 4) are idioms of war.

The reader gets the first mention of the state as actors, as the "headlights of the military police patrol car (…) were joined by officers from Thames Valley force, begin to climb the mound in pairs to bring down the protesting women" (Smith, 1983, p. 4). A second narrative of meta-journalism—the journalists becoming actors in the narrative—is also a reoccurring theme. Bishop (1983) writes, "The women climbed into the camp with embarrassing ease, watched by photographers, reporters and TV cameras" (p. 2). Furthermore, Bishop links the media coverage of Embrace the Base to an increase in women at the Peace camp for the Dancing on the Silos action, as "many of the demonstrators yesterday were said to be recent converts to the cause (…) after the mass demonstration last month" (1983, p. 2). At the Greenham Common Peace Camp, some journalists left their media work to join the camp, but many stayed for a week or two after the big mobilizations.

Media coverage was a double-edged sword, by generating the desired publicity that was missing from the early mobilizations, but it also changed the mood of the camp for the second mobilizations, Dancing on the Silo's. A returner and former CND activists, B recalls how "The big protests did foster a sense of unity amongst the women, with women who'd come all from all over the country and it was all rather jolly you know" (interview with Greenham Common activists B, 2010). However, such unity was often short lived an once the media and the visitors left the camps, "there was a serious difference between the feeling at a big demo at Greenham, and then a couple of days later, there was a very different feeling amongst the, women in small groups" (interview with Greenham Common activists B, 2010).

The media's continual use of frames of metaphors and cliché has dominated the Embrace the Base coverage. The tone changes to one that makes greater use of metaphors of war and a narrative that includes political and state actors. Like the early reporting of Dancing on the Silo, six activists are quoted from the Embrace the Base action. However, over time the order of discourse moves from the women only, to include observations on media discourse and the police, although the women remain the primary definers.

Media discourse puts the women at the top of the hierarchy of credibility in the discourse. The language is upbeat and positive; actors are primary definers in the hierarchy of credibility. The shift in mood with the second action, could account for the change in reporting style of the second event, revealing tensions within the camp. As Chapter Five and Six discuss, a very similar result of the climate camps created the same problems for the movement from 2007 to 2010. Another case where splits and tensions arose from media coverage was the anti-roads movement (1990s) and the ubiquitous Swampy.

Prior to 1997, protesters and activists are present in news reports. The hierarchical order of discourse runs from activists, to media and the state. In the Greenham Common reports, define the activists generically, with many referred to as "women" and "protesters", and "spokeswomen for the group said the action was designed to draw attention (…) construction of the silos had been speeded up" (Bishop, 1983). Metaphors indicate caring, mother earth, and feminism. Journalists use metaphors of motherhood, the Women's

Institute (Greenham reports). Thus, the early reporting of the Greenham Common peace camp applied terms of conflict and war. Nevertheless, the women were primary definers through their direct quotes. However, by the second action—Dancing on the Silo's journalists define the women in normalising, generic terms—"a mother of two" (Palmer, 1982), "Two women of 73 and 63" (Bishop, 1983), "44 women" (Palmer,1982), and "peace women" (Bishop, 1983). The next example, Swampy, identifies a shift away from normalization and more towards individualism through the trope of Swampy.

The Voice of Radical Activists

In the 1990s, a growth in environmental activism from the likes of professional ENGOs such as Greenpeace and the more radical collectives of the anti-roads protests, anti-nuclear and class wars meant environmental media stories were high on mainstream media news agendas. At the start of the 1990s, journalists would take ready to air news packages from ENGO such as Greenpeace. Greenpeace Internationals' 24-hour newsroom fed journalists with live action footage of direct action. As Greenpeace took full advantage of new technologies journalists and television stations unquestionably lapped up the footage of direct action and civil disobedience (Anderson, 1997). Yet, after the Brent Spa campaign, where Greenpeace mistakenly estimated the least environmentally damaging way for Royal Shell Oil Company to dismantle an abandoned rig, led to a dispute between the German Chancellor and British Prime Minister over environmental responsibility, traditional media were less trusting of ready-made news packages from ENGOs.

Greenpeace seduced the broadcast journalists and television executives by their professional delivery of ready to air material for the media. Only after Greenpeace's Brent Spa campaign did Greenpeace retort against the media, claiming their willingness to take footage of activists on top of the rig, or their own footage were to blame,—"The media got drunk on the drama of the Brent Spar story and now they blame us for the hangover" (The Independent cited in Rose, 1998, p. 161). Journalists struck out at Greenpeace's mistake, and the backlash led to a break down in trust between NGOs, activists, and journalists. Rose (1998) singles out the BBC as failing to "testing any claim Greenpeace makes, it tends to raise the issue of the Spar to cast any doubt on a claim or a statement without testing it. It is as if the BBC feels it has to make up for transmitting the footage of the Spar campaign by repeatedly attacking Greenpeace for it" (pp. 164–165).

Journalists' eagerness to take the ready to air packages without questioning the contents or finding a counter position show how mixing the drama of environmental activism with poor media practice failed to question Greenpeace's position. Editors began pulling back from ready-to air news packages, but were still hungry for the drama of direct action and they found sustenance in the more radical ecoActivism movements of the anti-roads, free-music rave scene. The new novelty protest camps provided the drama of direct action, but as a non-professional group, they were easier for journalists to access without relying on ready-made professional ENGOs for footage. Reacting to increasing professionalization of environmental activism from ENGOs and a general distrust of mainstream political parties' neoliberal agenda to addressing climate change, numerous environmentally conscious individuals and collectives began to gain momentum through protest camps and reclaiming the cityscapes from big business. Although ecoActivists had been capacity building since the 1970s, there was a big expansion in the 1990s.

Reclaim the Streets; EcoActivism and Media Governance in 1990s Britain

The free music or rave scene were an "ecstatic sharing made possible by machines" (Gaillot, 1999, p. 33), and machines were malleable as agents of technology. The Rave scene embracing of technology meant change it "from an instrument of isolation and reduction" into an "instrument for bringing people closer together and for the free artistic political expression of their singularity" (Gaillot, 1999, p. 19. cf St John, 2003, p. 68). A movement of New Age Travellers, Hippies, Acid House DJ's, DiY politics, and a reclaiming abandoned or disused symbols of capitalism for pleasure. As St John (2003) eloquently puts it, "the small hours of the twenty-first century saw a groundswell of proactive interventions undertaken by the politically dispossessed and agitated (including multitudes of young people) in the name of democracy, the environment, fair trade, debt cancellation and so forth". St John adds that

> One of the most significance was the massive global street party held on 18 June 1999 (coinciding with the G-8 meeting in Cologne, Germany). On that day, there were actions in seventy cities around the world. Groups, including Reclaim the Streets and the Peoples' Global Action (along with Bangladeshi female garment workers and Indian farmers), blockaded stock exchanges, superstores and big brand outlets, and

> occupied the offices of transnational headquarters and business districts. Signalling the "crisis of legitimacy" of neo-liberal political agendas, J18 was the globalization of protest. (St John in Muggleton and Weinzierl, 2003, p. 69)

The rave scene provided a house share style relationship within counter culture, united in finding ways in which to challenge state and corporate power. Rave gave ecoActivists enough support to develop its own identity through the anti-roads protests. It was a period famously labelled DiY Protest (McKay, 1999). Reclaim the Streets, a reincarnation of Earth First! Movement in the USA began to protest by closing off the streets of central London to reclaim the space from cars. Cars, transport hubs, new road building projects, and airport expansion were all targets of the emerging ecoActivists movement.

After the Brent Spar saga, journalists were still hungry for the drama of ecoActivists and turned to the radical non-hierarchical environmental activists groups of the anti-roads and rave scene collectives. EcoActivists utilized media appearances to make public and physical statements on the environmental impacts of large-scale infrastructure projects and the increasing conflating of economics with environmentalism. As the decade raced towards the new millennium, a flurry of anti-roads and anti-aviation protests occurs between 1996 and 1999, and a tipping point for the UK ecoActivism movement. April saw Reclaim the Streets held a Festival of Resistance link up with sacked workers from a Liverpool dockyard against job cuts to the industry.

By 1996, the media and politicians were using the trope of one individual activist to represent the many environmental collectives. Swampy aka Daniel Hooper was an activist first at the A30 bypass protest camp (1996), and then Manchester Airport (1997). Swampy a nickname given to Hooper by the camp first came to the media's attention for living underground in a tunnel he'd "dug in" at a protest camp. The camp was set up to protest at the A30, a new road bypassing the Devonshire village of Fairmile, between the town of Honiton and the city of Exeter.

Tree camps, tunneling (or digging-in) was a popular protest tactic in the 1990s. Activists would build labyrinths of tunnels making the ground so unstable that large machinery, such as diggers were unable to enter the site without risking harm to the protesters below ground. At the A30 protest in Devon (1996) tunnelling at protest camps "reached a whole new level of ingenuity and dedication at the A30 protest" (Aufheben in McKay, 1998, p. 125).

Figure 2.1. Anti-roads protesters being evicted from Tothill camp, Newbury bypass. Feb 1996. Image courtesy of Nick Woolley (wikimedia).

Fellow activists invented the name Swampy, as he tells the press, "Friends gave me the name Swampy, and I use to like it. We all had nicknames back then in the protesting days but I don't use it any more or tell anyone". Swampy's popularity grew the longer he stayed underground. For journalists, digging in was a good way to recreate the tension and drama missing since the Brent Spar saga. With the spotlight on Swampy, and a lack of gatekeeping that was common with the ENGOs, journalists freely interviewed many of the activists. As we will see this lack of gatekeeping meant guaranteeing activists' voice in the media, but also no control in how the protest was reported.

National newspapers ran neutral headlines. *The Independent* exclaims "Risk grows as Swampy digs deeper" (Jury and Schoon, 1997); "One week on, Swampy comes out blinking" (Jury, 1997), "First it was tunnels, now its runways" (Aitkenhead, 1997). The Times newspaper's declares, "Last A30 tunneller emerges defiant" (Fresco, 1997). The publicity around the A30, meant the media were keen to follow Swampy and headed north with him to a protest in south Manchester.

Activists and local community groups had set up a camp protesting against the expansion of Manchester Airport, in northeast England in the Bollin Valley, Cheshire. Local residents had for years been against the airport expansion plans before the activists moved in. Manchester airports' second runway was first proposed in 1991, and approved in March 1993. However, unlike the Greenham Common media discourse two changes occur in these examples. Firstly, a lot of focus falls on a couple of individuals (Swampy, Ian, and Animal), and the activists' role as primary definer is increasingly replaced by voices of state authority. The shifting in the reporting within the boundaries of representation (Fairclough's term) is beginning to drift away from the activists and towards state representatives as primary definers.

The Independent newspaper replaces the activist as the primary definer with a quote from Undersheriff Trevor Coleman "obviously a matter of great concern" (Jury and Schoon, 1997, p 1). The metaphor of war returns, "fellow protester, Alan, an ex-Para, who fought with Swampy to prevent the new runway" (Ball, 1997, p. 3). Aside from Swampy, activists are nameless, secondary sources, described as "people who have led troubled lives and for the most part the sub-culture of the camp is their only family" (Fordham in Ball, 1997, p. 3), or "new age eco-warrior" (Ball, 1997, p. 3).

Discord in the camp is also highlighted when Swampy's celebrity status is linked to splits in the camps. *The Independent* newspaper note that "the surge of publicity that turned Swampy the environmental protester into a media celebrity is now creating waves of unrest among underground supporters" (Ball, 1997, p. 3). Whilst, the *Sunday Mirror* reiterates the trope of Swampy a celebrity, "Swampy's camp is buzzing with rumours he's about to be signed up by Hollywood to work on a film of his life story" (*Sunday Mirror*, 1997), the rationale for the protest is lost. The camp changes from a protest camp to "Swampy's camp" (ibid.).

Such divide and conquer reporting is a common tactic of the right-leaning press aiming to cause tension in the camps. As Swampy's publicity grew when national newspapers profiled him in the lifestyle pages, and the national newspaper, the *Sunday Mirror* gave Swampy a six-week column. Swampy was becoming the poster boy for a movement.

Swampy's Newspaper Column Undermines the Movement

The *Sunday Mirror* newspaper gave Swampy a weekly column in which to document the protest camp near Manchester airport. Swampy's column was constructed from short paragraphs, each a thought on the day's events, The columns were in the form of a weekly diary over a six-week period (March 9th to April 20th, 1997). The tag line "Britain's best-known eco-warrior writes for the Sunday Mirror: You really dig our boy!" (*Sunday Mirror*, August, 1997). The *Sunday Mirror* proclaims, "He [Swampy] has chosen the Sunday Mirror to keep the country informed with a column detailing the operation" (Roberts, 1997, p. 9).

The first column documents the days building the protest camp, digging tunnels, trips to court, and media coverage of the Manchester campaign. Swampy welcomes interview requests from national breakfast program, Good Morning Television (GMTV)—and says, "the attention is welcome so we can discuss the issues and the environmental damage done in the name of progress" (Swampy, 1997, p. 9). Over the next five weeks Swampy and fellow protester, Ian, from the A30 campaign, describe digging tunnels, building tree houses, as he and fellow activists defend the camp.

Progressively, the writing tone moves from jovial documenting of everyday camp life, to clashes with police and disillusionment with the political system. The first few columns are similar to a summer camping trip yet over time as the campaign gets more confrontational with local law enforcement, the tone shifts to one of antagonism and potential police brutality. In the fourth column, Swampy writes that;

> Monday, there was a lot of confusion about what went on last night. Demonstrators and police came to blows. I didn't see any of it because I was down a hole, but I hear that a policeman tried to arrest Martin and it turned nasty. Three of out of five people were injured—Martin had two broken ribs. It is hard not to get involved when you see a friend in trouble, but some of the demonstrators may have gone over the top. Non-violence must be the priority. (Swampy, March 30th, 1997, p. 11)

SUNDAY MIRROR, March 16, 1997 PAG

Britain's best-known eco-warrior writes for the Sunday Mirror

THE WORLD according to SWAMPY

You really dig our boy!

HE'S Swampy, Britain's top eco-warrior and our latest columnist. And you've all taken him to your hearts since we started publishing his diary written UNDER Manchester Airport.

Now Swampy will also have your letters and your gifts — everything from twists of tobacco for his roll-ups to food for his dog Bandana — to keep him digging.

By five to one, our readers wrote in backing his crusade to stop acres of countryside disappearing under the airport's second runway.

Janet Carroll, of Macclesfield, spoke for many when she described Swampy — real name Daniel Hooper — as "a hero of our times." She adds: "It's sad how vandalism can be called progress. Keep on digging."

Miss S.J. Dale, of Preston, adds: "Your love for life is a guiding light that I am determined to follow."

It wasn't only the young who backed Swampy. And one pensioner wrote to support his campaign sending all the money she could spare...a £1 coin.

S. Stockton, of York, says: "I think people like you are wonderful, more sensible and caring than those of my generation."

Pensioner *Mrs J. F. Tyndall, of South-East London*, says: "I would be proud to have a grandson like you someone who cares enough to do something about the environment. It will be your world one day and you are right to fight for it."

Mr M. Parfitt, from Gillingham, Kent writes: "How much more can we take? Our island is being lost to concrete, asphalt and the car."

David Brown, of Blackheath, South-East London, says it should be "Arise Sir Swampy." He writes: "You are brilliant and should be knighted."

Whatever you think, read on to find out what Swampy has been up to in the last week...

SUNDAY

TODAY I didn't progress much with my tunnel.

My mother visited with my sister Jo and we went for a pub lunch. I had chips and salad and a lager — vegan lager. The afternoon was taken up by a meeting.

We are not slackers. We are working hard to stop the destruction of the countryside and stop global warming.

We have a Kango hammer on hire, but it has to go back tomorrow.

MONDAY

THE weekend's over. It's a shame but most people disappear during the week. Site life is quiet. The Kango went back it cost us an extra £20 as it was clogged with dust. We used it to dig my tunnel.

Police helicopter circled camp for four hours. What a waste of taxpayers' money.

TUESDAY

IAN, who was in the tunnel with me at Fairmile, Devon, came down today. He is an excellent digger because he's got big muscles.

The tunnel is progressing fast. Ian is trying to be vegan — something everybody who cares about the enviroment should consider.

WEDNESDAY

THE contractors started work today putting up fencing, even though they promised not to until March 27.

We sat on the digger for several hours talking with police as to whether we should get off. I don't think they have quite got the grasp of it. Spent afternoon digging out start of chamber for more comfortable sleeping underground.

In the evening I spent some time with Merry. Sat and talked on pallet talked about life, not seen each other for a few days.

Merry is my partner, we have been together a few weeks. I met her here, she has done animal protests before.

She is a very strong person and a good laugh. We have similar views on life.

THURSDAY

THIS morning we fixed up a fan for the tunnel as air started getting short last night. Ian and I both woke up with headaches. The chamber is progressing at a good pace now. In the evening I got a phone call on the site mobile to inform me that some Tory says he would like to bury me in concrete. Bit of a mess up on his part.

I would love him to come down here and have a chat about it. He is being a fool, but he is a politician.

These are the people who are supposed to be running our country! I mean, he is only human and this is the point: No one should run for power because it breeds corruption.

We should be working with the land in small communities not working against the land, for large corporations.

I think we should not have leaders, everyone should make their own decision.

FRIDAY

ACTION still continues and site life still goes on.

As I am writing this I sit half-way up one of the banks near our camp. The birds are singing, there is a gentle breese and the river is flowing by

"VROOM," my silence is broken by a Boeing 747 polluter.

Another group of holidaymakers going to ruin another beautiful landscape with hotels, pools and discos as planes chuck out fumes devastating to the environment.

SATURDAY

DAVE is working on his tree house. This means we will have five at Sir Cliff Richard Camp. I must tidy mine and Merry's out later.

"I don't follow television, it is not relevant to our lives. I don't really listen to radio, either. Our relaxing time is having a chat around the fire in the evening. That is a nice evening for me.

Ian and I continue to work on the tunnel as the usual weekend influx arrive — other protesters and people who just want to chat.

It is good to see them. Much work needs doing. I wish we had this number mid-week, for then we could branch off over more of the routes.

If everyone who is thinking of coming down came even for a day or two there would never be a second runway.

We can live with that hope.

Let Swampy know what YOU think...write to him at Sunday Mirror, Canary Wharf, London E14 5AP

Figure 2.2. Swampy's column from the *Sunday Mirror* newspaper in 1997. Image courtesy of Mirrorpix.

Cementing his popularity and celebrity status Swampy's final column follows his appearance on a BBC TV news quiz—Have I Got News for You?, a satirical look at the last seven days. Swampy no longer charts everyday life of the camp, his friends, or the reason behind the action. Instead, there is a usurping of the daily protest life with a series of short opinion pieces from fox-hunting to meat eating, racism and the forthcoming general election.

There are three identifiable themes in the six columns; (1) a self-mocking of his "celebrity status", (2) a symbiotic relationship with local residents (this narrative reoccurs at the Heathrow Climate Camp and is discussed later in the book); (3) and a constant justification for direct action and protest camps. The column reads, "we [protesters] met lots of people who would be badly affected … many were very middle class but have realised the democratic system has failed" (March 16th, 1997).

It is worth noting that although Swampy's voice is heard, many of the words are juxtaposed with contradictions. Fifty percent of the columns are typeset next to advertisements for cheap air flights. The final column has an advertisement sponsored by the very same newspaper, the *Sunday Mirror* offering readers, return flights between the UK and Europe for £25.

The name Swampy became a synonymous for many future ecoActivism protests, a "byword for environmental non-violent direct action and youth disaffection from formal politics, often being used in headlines where he, as a person, never appears in articles" (Paterson in Seel, Paterson and Doherty, 2000, p. 51). Daily Mail newspaper's report on on the Occupy protest (2011) near the London Stock Exchange carries the headline "Move over Swampy, it's us who should be protesting" (Littlejohn, 2011). The Camp for Climate Action, aka Heathrow Climate Camp links back to Swampy. The *Independent* newspaper (Milmo and Bowcott, 2008) mentions Swampy at the Camp for Climate Action (Heathrow, 2007) as "new wave protesters target airport expansion: University educated campaigners are learning lessons from Swampy and the 1990s road protests". Hooper's presence at the Heathrow Camp for Climate Action (2007) reports "Swampy joins airport camp: (but goes to sleep at his mum's)" (Hodgson, 2007, p. 21). The *Daily Mirror* says "The return of Swampy underground eco-hero joins the Heathrow protest" (Owen, 2007). In reality, Hooper's presence at the Heathrow climate camp caused similar splits and friction within the movement as with the Manchester Airport action. As a Climate camp activist notes, "The Standard [London Evening Standard newspaper] actually interviewed Swampy about Heathrow" (interview with activist Mick, 2011). Another stalwart activist,

Steve (2011), says, "(...) he was on television. That was embarrassing". (Interview with activist Steve, 2011)

Boundary shifting, order of discourse and use of metaphors remain current in the reporting as authority shifts away from other ecoActivists. The self- and outwardly mocking in the columns and focus on one individual removes any discursive space of political debate (Jordan, 2002). The Protest camps just become empty vessels as the focus remains on Swampy's actions. Language is less authoritative with casual colloquialism compared to the police and state representations. In the *Independent newspaper* stories, the first direct quote from Swampy once out of the tunnel, is "I'm alright—I had plenty to eat" (Jury, 1997). A second ecoActivist, Ian, also, "agreed to come out if he was given a cigarette, cup of coffee and a chance to speak to the press" (Jury, 1997). The direct quote gives a voice, but the informal language and lack of reference to the protest is missing from the narrative.

Comparing this sample with the reporting of Greenham Common Peace Camp, there are clear differences in the media strategies of each movement. Despite the internal divisions, the Greenham Common Peace camp media strategy was very much about one voice of the camp, the collective voice. The peace camp has a media strategy summed up with one slogan "there are no leaders here, all the stars are in the sky" (Abraham, 1988). Greenham Common Peace Camp mantra was to, a) prevent individuals from being the focus, and b) have a small handful of voices speaking on behalf of a very diverse space. Whereas media reporting of the anti-roads protest was heavily skewed towards Swampy and a handful of individuals speaking on behalf of a movement.

A lack in the day-to-day detail from both protests camps makes the reportage romantic. The column focus on conflict as a discourse of violence starts to emerge. Swampy does retain some hierarchy of credibility, although there is a decrease in the boundaries of representation away from ecoActivists towards police, governments, and journalists who begin speaking on behalf of activists.

Swampy was the only time one individual spoke for the entire movement. From Swampy and until the Camps for Climate Action (2005–2010) (see Chapter Six), many activists spoke as collectives with one voice. A combination of direct, secondary and commentary from journalist, plus an identification of the actors places protesters as actors. The celebrity status put onto Swampy, detracts, but still keeps the protesters towards the top of the hierarchy of credibility. After the protest in Manchester the activist voice starts to be lost to the point by the turn of the century, the voice is almost silent.

May Day, May Day Media Governance

The end of the 1990s and the birth of a global internet, a change in political government in the UK and the expansion of the Global Justice Movement (GJS) saw ecoActivism aligning with the anti-capitalists rhetoric. A series of local, European and Global Social Forums, counter-summits and protests against government support for corporations, with poor environmental records and rapid expansion in growth supported by a neoliberal politics. Internationally, the North American Free Trade Agreement (NAFTA) a trade agreement between the USA, Mexico, and Canada saw large-scale protest against the corporatization and commercialization of water resources. Smith, Byrd, Reese, and Smythe (2015) found the "the Zapatista struggle found much sympathy among GJS groups across the globe, inspiring subsequent meetings in the same spirits in Spain in 1997 and Brazil in 1999" (2015, p. 13). Global Justice protest took place across Europe at large-scale political meeting such as the Group of Eight (G8) and Group of Twenty (G20), the United National Framework Convention on Climate Change (UNFCCC), Conference of the Parties most prominently in, Copenhagen (Cop 15, 2009) Denmark; Durban, South Africa (CoP17, 2011) and Paris, France (CoP 21, 2015) with each summit acting as conduits for protest.

In the UK, May Day protests (2000–2004) began as a combination of anti-capitalist and environmental protests as part of the global justice movement. Two main protests took place in the City of London, and around Trafalgar Square in central London. J18 is 18 June 1999; N30 is action on 30 November 1999, both occurring before the Seattle action. J18, N30 and the global event at Seattle (1999) introduces the global justice movement to the world's media. The street mobilizations in Seattle were "making politicizing claims about institutions considered their adversaries, such as the WTO or the International Monetary Fund (IMF)" (Smith *et al.*, 2015, p. 50). The protest aimed to spotlight global inequalities between richer and poor nations, and harvesting the environmental for economic gains. Between 2000 and 2004, the traditional May Day Marches of the socialist Labour Day unions and workers also saw anti-capitalist protest in cities around the world.

The Times, *Independent* and *Daily Mail* newspapers all ran front page stories for the May Day 2000. "Guerrilla gardening" activists, taking the lead from the earlier Reclaim the Streets movement planned to turn public urban spaces into green havens. The protest march began with a Critical Mass cycle ride, meeting up with other marchers outside the Palace of Westminster, in

Parliament Square, Westminster, London. Establishing a temporary camp on Parliament Green, tents were set up, people planted flowers and other greenery (including marijuana plants) in the flowerbeds, as activists dug up turf from Parliament Green and placed it onto the road. A slice of turf was placed in a Mohican style across the statue head of former Prime Minister, Winston Churchill. This act of culture jamming—"a generic name given to a range of activities which seek to re-work and reconceptualises elements from mainstream culture in order to make some kind of satirical comment" (Gilbert, 2008, p. 96), was the focus of the media imagery to frame the action.

Churchill's culture jammed Mohican headwear drew cries of outrage from the politicians and the press as "Riot yobs desecrate Churchill monument" (Sullivan, 2000, p. 1) and "Rioters shame legacy of Churchill" (Black, Dixon, Mitchell, and Swift, 2000, pp. 1–3). Moral outrage frames *The Sun* newspaper, with a call to "Find these animals", as a "Riot rocks London" (Parker, Sullivan, and Whitaker, 2000, pp. 4–5). Activists are now "mobs" who "picked establishment symbols and McDonald's for its rampage" (Parker *et al.*, 2000, pp. 4–5) or were as *Guardian* journalist John Vidal defines the activists as a "reasonably behaved mob looking for a catalyst" (2000, p. 3). There is similar reporting over the following years (2001–2004), and there emerges a discourse of security and protest as a security threat in the traditional media coverage.

Meade's (2008) study Mayday, Mayday! Newspaper framing anti-globalizers! On the global justice protest in Dublin, Eire (2004), found that newspaper "articles ultimately bolstered the dominant frame of anti-globalization as a security problem, suggesting that without a heavy police presence such events would not be peaceful" (2008, p. 343). The *Irish Independent* newspaper provided a highly charged inferential structure that anticipated violence. Readers were invited to adopt "the subject positions of the security forces or at least the security conscious, repeated discussions of the scale of the policing operation emphasizing the gravity of the impending crisis". Meade's findings are similar to the reporting of the 2001–2004 May Day protests in London, and a sense of the police controlling the protest under the rubric of a security problem.

The *Daily Mirror* claims "4,000 police tame anarchists", and protesters "kettled" by the police at Oxford Circus. This reflexivity measures by the state and informing public opinion a (technique of dominance in Foucault's terms); show how the hierarchy of credibility and order of discourse has now moved from ecoActivists to state driven by a narrative of security. Security begins to be a rationale for state mobilizations and a dominant theme in future reporting of any forms of ecoActivism. Security or the threat off breaches in security frames the narrative of future action involving global justice /anti-capitalist and later other climate protest camps.

Lee, Peachy, Urquhart, and Tendler (2000) note the organizational element of the state, when "After nearly six months of planning Scotland Yard and the City of London force had 5,500 officers ready on the streets for trouble (…) and another 9,000 in reserve across London". Meaning is shaped from official sources, as "the Yard had intelligence that a hard core of several hundred protesters was bent on using the demonstration as a cover for hit and run attacks" (2000, p. 3). As Boggan, Bennetto, Milmo, and Beard (2001) say, "London was braced for riots and looting". The first half of the article challenges the pre-emptive assumption of violence, yet the second half focuses on violence when:

> At Oxford Circus shortly before 3 p.m., it was here, if anywhere, that trouble was expected … a dark green R registered Jaguar car was damaged as protesters climbed on the roof and kicked the front bumper off the vehicle. BUT, there was no obvious damage to shops, many of which had been boarded up ahead of the demonstrations. (Boggan *et al.*, 2001, p. 4)

There is also a quote from an activist as 'one protester, who gave her name as Spirit, said "this was always intended to be a peaceful day. The police provoked what little violence there was by their very presence"' (Boggan *et al.* 2001).

At the G20 Meltdown in the City London action (2009), an amplification of fear merges all forms of ecoActivism into wider social issues. Journalistic language connects together (or labels) all the collectives (i.e. climate camps, environmentalists, anti-capitalists) with other (often more violent) collectives, to "escalate" (Hall, 1978, p. 223) a security threat. Journalists identify a specific issue and link it with a "subversive minority" (Hall, 1978, p. 223). Journalists often meld ecoActivists, the Whitechapel Anarchist, and Black bloc into one single group. The melding of the very different groups results in making it the normalized position of the reporting one of anarchism and extremism. *The Times* newspaper reports "Hospitals all set for victims of G20 violence; London is braced for riots in City streets as protesters vent anger" (O'Neill, 2009). Adjectives used to define the march included "alert" and "injured", with protesters expected to "storm buildings". Similarly, the Guardian's phrases defining the day included "pandemonium", "anarchist cells" and "resurgent anarchists" (Guardian, 2009, p. 4). Therefore, the framing of ecoActivism as a security threat becomes the common reporting position for journalists, reiterated by politicians and corporations. However, by 2007 the Camp for Climate Action began contesting this security discourse through the courts and professional bodies.

Camp for Climate Action 2007

The 2007 Camp for Climate Action was a both a pinnacle and turning point for the relationship between journalists and protesters. The 2007 Camp for Climate Action was the third (of five in total) approximately ten day pop-up protest camps, more commonly known as the Climate Camps. Heathrow was an environmental activist protest against the proposed expansion of London's Heathrow Airport. British Airports Authority (BAA) Heathrow owners were seeking government planning permission to build a third runway. A collection of different environmental groups (Climate Rush, So We Stand, Rising Tide, Camp for Climate Action) and NGOs (including HACAN[4] NOT RAG, and FTF), along with local residents from Sipson and Harlington villages. The two villages, Sipson and Harlington, would have to be demolished to make way for the third runway. Unified against the expansion the villagers converged with activists' collectives in protest against the plans. The outcome was a weeklong environmental protest camp against the proposal in a field near the airport.

The Sun's headline "Camp Crustie" (Francis, 2007, p. 20) defines activists as "irate", "hippy crusties" and "strangers to soap". All these terms we commonly used metaphors, definers, and cultural codes drawn from reporting of the anti-roads movement. There is no direct quote or primary source from protesters, as readers are told, "The worst thing you can do is make eye contact with one—they'll have you doing jobs in no time". When a protester is mentioned, it is as "Claire, but who looked more like a rainbow or babbling brook" (Francis, 2007, p. 20). The journalist uses the established language to reiterate a discourse that has become 'true' only through repeated use of signification spirals, and removal of activists as primary definers. In Foucauldian terms this is a "manifestation of truth" (1991, p. 8) or Fake news to use the current parlance.

Such language shapes a discourse as being *true*, a common understanding based on objective journalism to the point that even the protesters don't know for sure if it's a truth of not. A security discourse is present as "extremists" who will "hijack" the camp and "infiltrate the terminal buildings by posing as passengers". The security discourse is prominent in the *Daily Telegraph* and *The Sun* newspapers as "Anarchists plotting to disrupt flights" proclaims the *Daily*

Telegraph (Milward, 2007, p. 1) by "disguising themselves as holiday makers to cause havoc", "planning bomb hoaxes", "assaulting Heathrow's perimeter fence", and "boarding planes and then refusing to take their seats" (Milward, 2007). The *London Evening Standard* reporting takes a similar discourse under the metaphor of security and threat, as the sub headline reads "hoax packages to cause security alerts", as activists plan an "assault on airport fences and shops" (Mendick, 2007). The journalist replaces the activist's voice and identity with the terms "militant", "Eco-warriors", and "anarchists" (Rosser, 2007). A third article, "Eco-protesters target school-run mothers" (Rosser, 2008, p. 17) makes claims that protesters plan to "deflate tyres of 4×4s in areas such as Kensington and Chelsea, by jamming the valves open with mung beans" (Rosser, 2008, p. 17), a claim challenged by the Camp for Climate Action.

A landmark case came out of *London Evening Standards* reporting of the camp, when a collective decision at the Heathrow camp lodged a formal complaint against the *London Evening Standard* with the Press Complaints Commission (PCC) (see Chapter Five). Thus increasing use of security—framed reporting and the reversal of the boundaries of representation all contributed to Greenham Common Peace Camp, the anti-roads including Reclaim the Streets, and the May Day Protest all slowly disbanded to reform as engagement with the press become a point of contention with movements. However, with RTS there is a clear case of state intervention forcing the movement to disband. One stalwart activists recalls,

> It didn't have to do with the press, it had to do with the fact that The City and certain individuals were told by Central Intelligence that if they didn't want to be sued for £2.5 m by The City, then RTS was to disband … and then there was "Reclaim the Future." (Interview with activists MC, 2011)

Another former RTS stalwart feels the movement lost focus a common problems as discussed in Chapter Six, and blaming press relations is an excuse, he says,

> I tend to think it was wider than the press block. I think also that the movement lost a bit of focus because it had focus on the anti-roads then it had focus on the streets, and then there wasn't too much focus and think it was wider than just the media. (interview with Paul, 2013)

Whilst another activists Steve thinks the problem of burn-out combined with state intervention and media reporting contributed to its end, as he says

> On RTS, there were lots of different splits that we'd live with for years. Eventually people just thought, when yet another schism occurred, and that was when everyone thought, Oh, no, I just can't take another one.

The continual disagreements were what MC saw as pressure from the state signals the beginning of the end for RTS "it was disbanded by the State, is my reading of it. A lot of the people who wanted to do environmental, climate change; social justice type stuff became London Rising Tide" (interview with MC, 2011)

Whatever the cause, there is a clear indication that the voice of the protester is replaced as the boundaries of representation moves toward politicians, and the law as primary definers and away from activists. Each of these critical discourse moments act as nodal communicative events that frame public understanding of environmental protest within environmental discourse. Over time, the production and consumption of understanding ecoActivism is bridged through historical events to provide journalistic traits to supports a certain political discourse creating new cultural codes and maps of meaning.

Subsequently, journalists fall back on the same terms and lexicon to report on ecoActivism. Swampy becomes a simile for environmental activists, politicians use negative terms to refer to any activism, and increasingly the news report frames protest and activism as violent and destructive. Press reports return to Swampy as a metaphor during the first London May Day action. Plows notes (2002) as "eco-warriors and Swampy's friends" are now "anarchist anti-capitalist". Protest voices are replaced by political statements and editorials. The *Times Newspaper* reports Prime Minister, "Tony Blair condemned the mindless thuggery of the ecoactivist" and that "vandalism was beyond contempt" (Lee *et al.*, 2000, p. 1). Some protesters graffitied [sic] on the cenotaph, in "an angry statement the Prime Minister said 'it is only because of the bravery and courage of our war dead that these idiots can live in a free country at all. The police must have full support of everybody in dealing effectively with them' (Lee *et al.*, 2000, p. 1).

Journalists allude to protesters injuring police, as the research found in the May Day protests. Injury to protesters is negated in the reporting. Police causalities are tallied up, as "nine officers injured", whilst the police had spent "nearly six months planning Scotland Yard and the City of London Force" (Lee *et al.*, 2000, p. 1). Counting continues at the third May Day action—"five police officers were injured" despite "officers dressed in riot gear" as "police charged demonstrators" (Nawar, 2002, p. 7).

Metropolitan Police Assistant Commissioner David Vaness adds to the authoritative voice, noting, "we have high visibility policing and it will in effect be 'in yer face' policing with strong evidence gathering" (Nawar, 2002, p. 7). Vaness's comments are used for a *Daily Mirror* headline "In yer face" (Nawar, 2002, p. 7) and *Daily Telegraph* headline—"Police win May Day face off" (Steele and Pook, 2002, p. 1). Over the four year life of the May Day protest, front page headlines from *The Times* newspaper articles are "100 held as thugs riot in London anarchy" (Lee *et al.*, 2000, p. 1); The Independent newspapers' headline of "Veggie burgers, militant cyclists and rain" (Boggan *et al.*, 2001, p. 1) is juxtaposed with *The Daily Mail* sees "4000 police tame May Day militants" (Allen, Craven, and Taylor, 2002). The *Daily Mirror* quotes an unidentifiable protester describing the event as saying "we've been walking around in circles for hours; the day doesn't seem to have any focus" (Nawar, 2002, p. 7).

Journalists' negative tone is heightened by conjunctive words to change the discourse. Fairclough notes the use of sentence connectors "however", "but", "nevertheless" or "meanwhile" as "markers of the ordering of voices", and in a sentence it "implicitly contrasts positive and negative sides" (1995, p. 82). In the Greenham news reports, journalists apply "but" to create a contrast between activists and local residents. It's a technique used to divide and rule the activists. For example, Palmer (1982) applies "but" to juxtapose the women at Greenham Common Peace Camp with local residents, noting "but, not everybody at Greenham yesterday was happy" (1982, p. 1). Whereas in the newsreports of the global justice and May Day protests reporting "but" was used to mark out the dominant police position against activists: "smoke bombs and sticks were thrown but police in riot gear refused to allow protesters in or out" (Boggan *et al.*, 2001, p. 1). "Nevertheless" is often applied to give authority to the state by journalists. Their use of the term, for example, in "Nevertheless, the wanton rampage anticipated—and criticized in advance by the Prime Minister, Tony Blair and the Mayor of London, Ken Livingston, were kept in check". This flags up the pre-empting of violence, a theme that repeatedly emerges, from the Greenham Common protest reports to G20.

Use of neutral terms in identifying activists nullifies their voice, as "demonstrators", "protesters" and "people in groups" or "groups of people" (Jury, 1997). There are generic references to the state as the "judicial system", "police," and "legal system" (Bishop, 1983; Palmer, 1982). There is a similar prioritization of activists in the Swampy reports. Jury's (1997) piece starts with Swampy and protesters. The first quote in the piece comes from Swampy, followed by media discourse, then state—"police", "PPP consortium—connect a privately funding and building the £75m road and the government repays the cost over the next ten years" (Jury, 1997). Post 1997 the order of discourse puts the state as the dominant discourse. There are "mounted police", "police" and "violence" (Boggan *et al.*, 2001, p. 1) but no mention of activists. Only the following year do activists clearly re-emerge in the order of discourse. Allen *et al.* (2002) define activists as "hard core protesters", "anarchists" and repeat Boggan *et al.*'s (2001) use of the term "militant" in the headline. Thus the order of discourse moves from activists being directly quoted to negating the activists' position, and when they are defined, it is in terms of deviancy. The role of the state increases post-1997 and around 2000–2001 there is little mention of activists in the order of discourse.

Conclusion

From the early protest camps of Greenham Common to the Climate camps of the G20, CoP21, there is a reversal in the boundaries of representation. Primary definers move from protesters to police, politicians, and social commentators who in turn command the order of discourse and hierarchy of credibility. Reporting removes the activists from the narrative and replaced with comments from official spokespersons who speak on or about, but not within movements. The cognitive process involved in news production and decoding frame shape public understanding of ecoActivism from one of warmth to one of a threat of national security, as Luke (1999) predicted. Historically, traditional media frame all forms of ecoActivism as one with the removal of the protester voice, replaced by the police and government within the climate debate. The ideas and concept of any ecoActivist is no longer heard in the traditional media, although as we will see in Chapters Five and Six, activists have found ways to circumnavigate the reporting discussed in this chapter. Before looking at what strategies ecoActivists movements have developed to ensure they retain a voice we need to look at why and how protest has moved from setting the news agenda to being the news agenda. To do so we will look return to wider environmental and climate debates with a focus on environmental politics, governance, and economics.

When an environmental activist's voice is heard the language is informal, colloquial and can be interpreted as depoliticizing the event. Journalists' focus on colloquial language detracts from the reason for the camp, by presenting a notion of casualness and disorganization, in contrast to the organizational language of the state. A shift post-1997 also places the activists away from a discourse of motherhood and caring and into one of conflict. From 2000 onwards, a metaphor around ritualism, male rituals, war, and conflict is common in the journalistic language. There is an increasing use of the term "militant" as in "militant cyclists and eco-warriors" (Cobain, 2001), and the *Independent newspaper* ran with a page-two headline "Veggie burgers, militant cyclists and rain" (Boggan *et al.*, 2001, p. 1). The *Daily Mail* declares that "May Day militants", "Soho sex workers", and a "hard core" all took part in the 2002 May Day. This "amplification", to borrow Hall's term (1978), places the environmental activists into a framework that links numerous different movements, all appropriated through a discourse of deviance. This use of "noun-like" terms to define activists is what Fairclough terms "nominalization". Despite the order of discourse altering from the late 1990s onwards, journalists' discussion of media discourse remains a constant in the order of discourse. In establishing the media order of discourse, it gives an indication of fixed or shifting boundaries, discursive practice within the media order of discourse, and relations between media order of discourse and "socially adjacent orders of discourse" (Fairclough, 1995, p. 205).

The removal of the activists' voice continues into the new millennium. By 2007, journalists are beginning to shape a certain history of the movement through language. There reaches a point that the public now understand protests as a particular domain of objectivity. Journalists continue to use the same metaphors and exemplars, to frame protest from an objective position formulated from cultural codes developed in media reports of Greenham Common Peace Camp and anti-roads movement. There is an additional narrative or security and terrorism.

The increased co-opting of climate change rhetoric into the mainstream media by political parties and corporations meant that ecoActivists and professional environmentalists were no longer the lone voice of the environment. Politicians and later multinational corporations put forward an alternative economic approach through a variety of green capitalist schema. In the next two chapters we will look at how elite politics and global institutions slow co-opting of environmentalism has been detrimental to ecoActivism.

Notes

1. Halloran *et al.*'s (1970) study categorized journalistic language into nouns and adjectives. These categories were then slimmed down into sub-categories of nouns which were either

"neutral", "specific descriptions" or "unfavourable". Adjectives were sub-categorized as "favourable", "neutral," and "unfavourable".

2. Of the thirteen articles on Greenham Common Peace Camp for this study, those by Brown, Express Staff Reporter, Pratt (1982), Steven and the anonymous piece in the *News of the World* newspaper (1983) were rejected, because they were either shorter than 200 words or appeared buried in the newspaper. There were two articles retained, *The Daily Mirror* newspaper's "Peace: the plea by 30,000 women" (Palmer, 1982) and the *Observer's* "Peace women's raid: 44 charged" (Bishop, 1983). These two articles were kept for analysis as they were on the front page and each covers the two actions.
3. Given the vast range of articles, this section will highlight a mere few, and as with the earlier examples, attention will focus on two stories as representative samples of the reporting as a whole. From the sample there were thirty-two articles across four years that focus on the anti-capitalist/global justice protest (2000–2004). Due to the volume of articles and length of time, the years 2000, 2001 and 2002 will be given priority. There were twenty-nine articles, with the highest percentage from years 2000, 2001 and 2002. As with the overall criteria, there was a second round rejecting any articles that did not appear on the front page or initial news section, were comments or editorials, or were predominantly images and lacked text. For example, *Daily Mail* (Nawar, 2002) newspapers carried full-page spread images of protesters clashing with police but limited text.
4. HACAN—Heathrow Association for the Control of Aircraft Noise—Clear Skies. HACAN—Clear Skies began in the 1960 as HACAN. Its aim is to represent the concerns of residents under flight paths. For more information, see HACAN (2011). No Third Runway Action Group—Heathrow Airport. For more information, see NOTRAG (2011).

References

Abraham, E. (1988). "We Have No Leaders Here All the Stars Are in the Sky", Vol. 1, Scope and Content. Leaflet by Elizabeth Abraham of Violet Gate/Woad Gate, Greenham Common Peace Camp, concerning feminist revolution. Papers of Jill Truman.

Aitkenhead, D. (1997). "First It was Tunnels, Now It's Runways", *Independent on Sunday*, 2 February, p. 5.

Allen, Craven, and Taylor (2002) 4,000 police tame May militants, *Daily Mail* (London), May 2, p. 8.

Anderson, A. (1997). *Media, Culture and the Environment*, London: Routledge.

Aufheben. (1998). "The Politics of Anti-road Struggle and the Struggles of Anti-road Politics: The Case of the No M11 Link Road Campaign", in G. McKay (ed.), *DIY Culture: Party and Protest in Nineties Britain*, London: Verso.

Ball, G. (1997). "Swampy, a Tunnel Star Undermined", *Independent on Sunday*, 8 June, p. 3.

Becker, H. (1967). "Whose Side are We on?", *Social Problems*, 239, pp. 240–247.

Bishop, P. (1983, January 2nd). Peace Women's Raid: 44 Charged. *The Observer*, p. 2.

Black, Dixon, Mitchell, and Swift. (2000). "Rioters Shame Legacy of Churchill", *The Daily Express*, May 2, pp. 1–3.

Boggan, S., Bennetto, J., Milmo, C., and Beard, M. (2001). "Violence mars May Day Protest", *Independent*, 2 May, pp. 1ff.

Boyce, T., and Lewis, J. (eds.) (2009). *Climate Change and the Media*, New York, NY: Peter Lang Publishers.

Boykoff, M. T and Boykoff, J. M. (2007). Climate change and journalistic norms: A case-study of US mass-media coverage. *Geoforum: Journal of Physical, Human and Regional Geosciences.*, 38(6), 1190–1204.

Boykoff, M. T., & Mansfield, M. (2008). 'ye olde hot aire': Reporting on human contributions to climate change in the UK tabloid press. *Environmental Research Letters*, 3(2), 024002. doi:10.1088/1748-9326/3/2/024002

Brown, P. (1982). "Scuffles as Women Seal of Base", *The Guardian* (London), December 14th, p. 1.

Carvalho, A. (2000). *Discourse Analysis and Media Texts: A Critical Reading of Analytical Tools*, paper presented at the 'International Conference on Logic and Methodology', RC 33 meeting (International Sociology Association), Köln, 3–6 October. Available at http://repositorium.sdum.uminho.pt/bitstream/1822/3137/3/acarvalho_Kolnpaper_20k00.pdf (accessed August 2010).

Cottle, S., and Lester, L (eds.). (2011). *Transnational Protests and the Media*, New York, NY: Peter Lang Publishers.

Dexter, L., and White, D. M. (1964). *The "Gate Keeper": A Case Study in the Selection of News*, London: Free Press of Glencoe. Available at http://folders.nottingham.edu.cn/staff/zlizrb/2008_IC_ICT/Resources/White_1964.pdf

Doyle, J. (2007). "Picturing the Clima(C)Tic: Greenpeace and the Representational Politics of Climate Change Communication", *Science as Culture*, 16(2), pp. 129–150.

—— (2009). "Climate Action and Environmental Activism: The Role of Environmental NGOs and Grassroots Movements in the Global Politics of Climate Change", in T. Boyce and J. Lewis (eds.), *Climate Change and the Media*, New York, NY: Peter Lang, pp. 103–116

—— (2016) Mediating Climate Change, London: Routledge, 2016.

Eldridge, J. (1995). "Breaching the Peace at Greenham Common", in *Glasgow Media Group Reader; Vol. 1: News Content, Language and Visuals*, London: Routledge, pp. 318–334.

Express Staff Reporter. (1982). "The Peace War: Russian TV Camera Roll as 30,000 Women Ring Missile Base in Antinuclear Protest", *Daily Express*, December 13th, p. 1.

Fairclough, N. (1995). *Media Discourse*, London: Edward Arnold.

Foucault, M. (1970). *The Order of Things: An Archaeology of the Human Sciences*, New York, NY and London: Tavistock.

—— (1991). "Governmentality". In: Burchell *et al.* (eds.) *The Foucault Effect*, Hemel Hempstead: Harvester Wheatsheaf, pp. 87–104.

(2008).*The Birth of Biopolitics. Lectures at the Collège de France, 1978–79*. Edited by M. Senellart and translated by G. Burchell, Basingstoke: Palgrave Macmillan.

Francis, N. (2007). "Camp Crusty: Undercover, Overtired and Glad to Get Out", *The Sun*, 15August, p. 20.

Frenzel, F., Feigenbaum, A., and McCurdy, P. (2014). "Protest Camps: An Emerging Field of Social Movement Research", *The Sociological Review*, 62(3), pp. 457–474. doi:10.1111/1467-954X.12111

Fresco, A. (1997). "Last A30 Tunneller Emerges Defiant", *The Times* (London), January 31, p. 4.

Gaillot, M. (1999). "Multiple Meaning: Techno, An Artistic and Political Laboratory of the Present", Translated by Warren Niesluchowski, Paris: Éditions dis Voir.

Gamson, W. A., Croteau, D., Hoynes, W., and Sasson, T. (1992). "Media Images and the Social Construction of Reality", *Annual Review of Sociology*, 18(1), pp. 373–393. doi:10.1146/annurev.so.18.080192.002105

Gamson, W. A., and Modigliani, A. (1989). "Media Discourse and Public Opinion on Nuclear Power: A Constructionist Approach", *American Journal of Sociology*, 95(1), pp. 1–37. doi:10.1086/229213

Gerbaudo, P. (2012). *Tweets and the Streets: Social Media and Contemporary Activism*, London: Pluto Press.

Gilbert, J. (2008). *Anticapitalism and Culture: Radical Theory and Popular Politics*, Oxford: Berg.

Gitlin, T. (1980). *The Whole World is Watching: Mass Media in the Making & Unmaking of The New Left*. Berkeley, CA: University of California Press.

Griggs S., and Howarth, D. (2000). *Discourse Theory and Political Analysis: Identities, Hegemonies and Social*, Manchester: Manchester University Press.

Hall, S. (1973). *Encoding and Decoding in the Television Discourse*, Birmingham: Birmingham Centre for Contemporary Cultural Studies, University of Birmingham.

Hall, S. (1997a). *Representation and the Media*. Hall, S., Jhally, S., Talreja, S., Patierno, M. and Media Education Foundation (Directors). (1997).[Video/DVD] Northampton, MA: Media Education Foundation.

Hall, S. (1997b). *Representation: Cultural Representations and Signifying Practices*, London: Sage Publications.

Hall, S. (1978). *Policing the Crisis: Mugging, the State and Law and Order*, London: Macmillan Press.

Halloran, J., Elliott, P., and Murdock, G. (1970). *Demonstrations and Communication: A Case Study*, Harmondsworth, Middlesex: Penguin.

Hansen, A. (1993). *Mass Media and Environmental Issues*, Leicester: Leicester University Press.

Hansen, A. (2010). *Environment, Media and Communication*, London: Routledge.

Hodgson, G. (2007). "Swampy Joins Airport Camp: (But Goes to Sleep at His Mum's)", *The Sunday Mirror*, 19 August. Available at http://www.thefreelibrary.com/swampy+joins+airport+camp%3b+exclusive+(but+he +goes+to+sleep+at+his-a0167745322

Hutchin, B., and Lester, L. (2006). "Environmental Protest and Tap-dancing with the Media in the Information Age", *Media, Culture & Society* SAGE Publications (London, Thousand Oaks and New Delhi), 28(3), pp. 433–451.

Jordan, G., and Maloney, W. (1997). *The Protest Business? Mobilising Campaign Groups*, Manchester: University of Manchester Press.

Jordan, T. (2002). *Activism! Direct Action, Hacktivism and the Future of Society*, London: Reaktion Books.

Jury, L., and Schoon, N (1997). "Risk Grows as Swampy Digs Deeper", *The Independent*, 30 January, p. 1.

Jury, L. (1997). "One Week on, Swampy Comes out Blinking into the Television Lights", *The Independent*, 31 January, p. 1.

Klein, N. (2014). *This Changes Everything: Capitalism vs. the Climate*, London: Allen Lane.

Lang, G. E., & Lang, K. (1955). "The Inferential Structure of Political Communications: A Study in Unwitting Bias", *The Public Opinion Quarterly*, 19(2), pp. 168–183. doi:10.1086/266559

Lee, A., Peachy, Urquhart, P., and Tendler. (2000). "100 Held as Thugs Riot in London Anarchy", *The Times*, 2 May, pp. 1–3.

Lester, L. (2007). *Giving Ground: Media and Environmental Conflict in Tasmania*, Tasmania: Quintus Publishing, UTAS.

Lester, L. (2010). *Media and Environment: Conflict, Politics and the News*, Cambridge: Polity.

Lester, L., and Hutchins,B. (2013). *Environmental Conflict and the Media*, New York: Peter Lang

Littlejohn, R. (2011). "Move Over Swampy, It's Us Who Should be Protesting", *Mail Online*, 18 October. Available at http://www.dailymail.co.uk/debate/article-2050223/Occupy-London-Stock-Exchange-Its-protesting.html (accessed January 2012).

Luke, T. (1999). "Environmentality as Green Governmentality", in E. Darier (ed.), *Discourses of the Environment*, Oxford: Blackwell, pp. 121–151.

McKay, G. (1998). *DiY culture: Party and protest in nineties' Britain / edited by George McKay*. London: Verso.

Meade, R. (2008). "Mayday, Mayday! Newspaper Framing Anti-Globalizers! A Critical Analysis of the Irish Independent's Anticipatory Coverage of the 'Day of the Welcomes' Demonstrations", *Journalism*, 9(3), pp. 330–352.

Mendick, R. (2007). "Extremist to Hijack Climate Camp Demo. Militants in Plot to Paralyse Heathrow", *London Evening Standard*, 13 August, p. 1.

Milmo, D., and Bowcott, O. (2008). "New Wave Protesters Target Airport Expansion", *The Guardian*, 1st March [online]. Available at http://www.guardian.co.uk/environment/2008/mar/01/activists.theairlineindustry (accessed 1 September, 2010)

Milward, D. (2007). "Anarchists Plotting to Disrupt Flights", *The Daily Telegraph* (London), August 14, p. 1.

Nawar, S. (2002). "In Yer Face", *Daily Mirror*, 2nd May, p. 7.

O'Neill, S. (2009b). "Hospital All Set for Victims of G20 Violence; London is Braced for Riots in City Streets as Protesters Vent Anger", *The Times*, 27 March, pp. 16–17.

Owen, G. (2007). "The Return of Swampy: Underground Eco-Hero Joins the Heathrow Protest", *Daily Mirror*, 19 August, page unknown.

Painter, J. (2013). *Climate Change in the Media: Reporting Risk and Uncertainty* (1st ed.). London: Published by I. B. Tauris & Co. Ltd in association with the Reuters Institute for the Study of Journalism, University of Oxford.

Palmer, J. (1982). "Peace: The Plea of 30,000 Women Who Joined Hands in World's Most Powerful Protest Against Nuclear War", *Daily Mirror*, 13 December, pp. 1–3.

Parker, N., Sullivan, M., and Whitaker, T. (2000). "Find These Animals", *The Sun*, 2 May, pp. 4–5.

Paterson, M. (2000). "Swampy Fever: Media Constructions and Direct Action Politics", in B. Seel, M. Paterson, and B. Doherty (eds.), *Direct Action in British Environmentalism*, London: Routledge, pp. 151–167.

Philips, L., and Jørgensen, M. (2004). *Discourse Analysis as Theory and Method*, London: Sage.
Plows, A. (2002). *Praxis and Practice: The 'What, How and Why' of the UK Environmental Direct Action Movement in the 1990s*, unpublished Ph.D. thesis, University of Wales, Bangor.
Pratt, C. (1982, p. 5). Peace War that Won't Go Away. *Daily Express*, p.5.
Roberts, D. (1997). "Britain's Best-Known Eco-Warrior Writes for the Sunday Mirror: You Really Dig Our Boy!", *Sunday Mirror*, 30 March, p. 9.
Rose, C. (1998). *The Turning of the Spar*, London: Greenpeace.
Rosser, N. (2007). "Eco Warriors Hit Heathrow", *London Evening Standard*, 24 May, pp. 1–2.
(2008, p. 17) 'Eco protesters target school-run mothers', *London Evening Standard* Newspaper, section b, March 31, p. 17.
Saunders, C. (2013). *Environmental Networks and Social Movement Theory*, S. l.: Bloomsbury Academic.
Smith, J., Byrd, S., Reese, E., and Smythe, E. (2015). *Handbook on World Social Forum Activism*, Boulder, CO: Paradigm Publishers.
Smith, J. (1983, December, 14th). Peace Women Go Over the Top. *The Sunday Times*, p. 2.
Steele, J., and Pook, S. (2002). "Police Win the May Day Face-off", *The Daily Telegraph* (London), May 2, p. 1.
St John, G. (1967–2003). "Post-Rave Technotribalism and the Carnival of Protest", in D. Muggleton and R. Weinzierl (eds.), *The Post-Subcultures Reader* (1st ed.), New York: Berg.
Sullivan, M. (2000). "Riot yobs desecrate Churchill monument", *The Sun*, 2 May, p. 1.
Sunday Mirror. (1997). "Swampy's Camp Buzzing with Rumours", *Sunday Mirror*, 30 March, p. 9.
The Guardian. (2009). "Police Tactics Queried as Met Says G20 Protests will be 'Very Violent'", *The Guardian*, 28 April, p. 1.
Van Dijk, T. (1988). *News Analysis: Case Studies of International and National News in the Press*, Hillsdale, NJ: Laurence Erlbaum.
Vidal, J. (2000). "Ben Threw the First Stone – and the Garden Became a Running Battle with Police", *The Guardian*, May, p. 3.

· 3 ·

ENVIRONMENTAL GOVERNANCE

The Role of Environmental Activism in Contentious Politics

> Most environmentalist movements now operate as a basic manifestation of governmentality.
>
> —Luke (1999, p. 22)

> In the Western capitalist countries, the first task of governments, in environmental policy, and everything else is to keep actual and potential corporate investors happy.
>
> —Dryzek (1997, p. 11)

Climate change shifts from an environmental issue to an increasingly political one during the 1990s and into the early millennial years. In the USA, UK, Australia, and other western countries, politicians were beginning to understand climate change as a public policy issue. The United Nations (UN) saw a need for a more global approach incorporating both the developed and developing countries of Africa, South America, Asia and the Pacific region. Four years after the Rio Earth Summit, the inaugural Conference of the Parties (CoP) meetings (est. 1995, Bonn, Germany) began. CoP was tasked with reviewing, "national communications and emission inventories submitted by Parties".[1]

The UN's Rio Earth Summit (1992), became a tipping point for global environmental action, with the formulation of Agenda 21 (A21). The summit was significant because it marked the turning point in green politics when the UN "adopt the first draft of the 'Earth Charter', a vision for an environmentally sustainable planet" (Cox, 2010, p. 78). The Charter put international pressure on large organizations to adopt an environmental position. The United Nation's Conference on Environment and Development (UNCED), passed a motion on Agenda 21, which "propelled and justified the term 'sustainable development' into common currency" (Buckingham and Turner, 2008, p. 50). Realizing that consequences of a highly industrialized global economy (construction, infrastructure developments) was having on the global climate, the UN set out A21 as a mandate for every country and corporations to include sustainability into large-scale projects and everyday living. The UK, USA and Australian mainstream political parties take a neoliberal approach to climate change solutions. New strategies and policies created new energy and alternative markets; with some believing the only solution to the problem was centered in economics. State governments, the United Nations (UN), the Intergovernmental Panel on Climate Change (IPCC) and even oil companies such as Beyond Petroleum (BP) were vying to become authoritative voices of environmental discourse by greening their products. Many large multinational corporations rebranded as green, famously BP Amaco changed to Beyond Petroleum accompanied by a green and yellow sunburst logo (2000). However, accusations of greenwashing and large-scale environmental disasters (such as Deepwater Horizon oil spill, 2010), manipulation of carbon emissions schemes (discussed later in this chapter), and eco-labels led some to question if creating a green economy was the right way to address climate change, or just capitalism through a green lens. The introduction of flexible mechanisms through the Kyoto Protocol (1996), and market-led initiatives epitomize what Timothy Luke calls "green govermentality" (1999) similarly, Oels (2005) terms the same process "eco-governmentality".

Green Governmentality: Managing the Environment through Persuasion and Discipline

With economics a key driver in governments' working with business over environmental concerns, Dryzek says, "in the Western capitalist countries, the first task of governments, in environmental policy, and everything else is to keep actual and potential corporate investors happy" (Dryzek, 1997, p. 11). Governments and corporations advocate for equality and just society through economic solutions, with many governments using the idea of governance and governmentality to balance corporate investor needs with those of the public. What is governance in an environmental context?

Governance is not government or the art of government but two poles on a continuum of different governing types (Finer, 1970). Bevir and Rhodes observe, "Governance is a change in the nature or meaning of government" (2003, p. 45), and this is expanded in Stoker's definition that governance refers to the emergence of "governing styles in which the boundaries between the public and private sectors are blurred" (1998, p. 17). Pierre and Peters (2000) note "the term governance derives from its capacity—unlike that of the narrower term—government—to cover the whole range of institutions and relationships involved in the process of governing" (Pierre and Peters in Jordan, Wurzel, and Zito, 2003, p. 478). Rhodes calls such governing styles a "new operating code" (1996, p. 47) that works as a "self-organising and co-ordinating network of societal actors" (Schout and Jordan, 2005 in *et al.*, 2005, p. 480). Despite these varying definitions, Jordan *et al.* (2003), echoing Schout and Jordan, offer a useful interpretation of governance that it is "synonymous with a change in the meaning of government, a new process of governing, or a changed condition of ordered rule, or a new method by which society is governed" (2005, p. 477).

A new green economy and the environment went hand in hand with government policies around the same time as the environmental movement began. Whereas ENGO and activist groups take an holistic approach to conservation and environmental management, governments tended to take administrative, governance position, embedded in an anthropocentric focus. In other words, governments began to administrate and regulate ways for people to be dominant over the environment—as opposed to a symbiotic, holistic approach favoured by ENGO and activist groups. For consecutive governments, the environment became a way to administrate, cajole, and convince people through promotion of sustainability and eco-friendly behaviour. In the USA, Al Gore's Global Marshall Plan was an economics driven solution based on the Marshall Plan that revived American imperialism after World War II (1992). In the UK, the environment became a thing to be managed and regulated through policy and practice.

This chapter explores how in the search for climate change solutions governments will try to persuade the population to seek solutions through consumer markets such as ecolabels, eco-vehicles, etc. This chapter unpacks how various governments use green governmentality to help persuade the population that economics is the best solution to the climate change problem. In doings so, the chapter draws on Luke's interpretation of Foucault's concept on governmentality through a green lens (green governmentality) to see how the general population are persuaded to consume their way out of the problem.

Was Foucault an Environmentalist?

Governments' realization there was a need to better manage the environment led to the creation of Department of Environments (DoE) in the UK, USA and Australia, amongst others throughout the 1970s. DoE's meant governments were able to ensure clean air, water and the population growth in cities and towns. By creating Department of Environments (DoE) it also allowed for easier administration and regulation of people. The administration of the population is for Foucault a form of biopower whose governance is administrated through biopolitics to affect behavioral change.

At its extreme biopower can take away all forms of life. The administration of life for example with the nuclear debate, or the "atomic situation" (Foucault, 1978, p. 137) is an analogy for power through administration. Nuclear weapons means governments have the power to extinguish entire populations.

> The power to expose a whole population to death is the underside of the power to guarantee an individual's continued existence … it is because power is situated and exercised at the level of life, the species, the race, and the large-scale phenomena of population. (Foucault, 1978, p. 137)

This atomic situation is a culmination, an ultimate show of the power of one person over another, with the ability to take the life of another—the ultimate form of control. In political terms, the analogy works to reveal an underlying principle (that one has to be capable of killing in order to go on living) as "at stake is the biological existence of a population, this is not because of the recent return of the ancient right to kill" (Foucault, 1978, p. 137). We can apply Foucault's atomic and nuclear questions to show how politics and control over human life can be linked together with a green governmentality approach to environmental governance.

Environmental governance first appears with the formation of national parks. Designating areas of the wilderness to manage and maintain was in effect the formalization of nature. The first national park was Yellow Stone in the USA (1872), Banff in Canada (1885), Australia Royal National Park (1879) and the Peak District, UK much later in 1951. In 1975, the Australian government creates the world's first Marine national park with the Great Barrier Reef Marine Park Act (1975). With less focus on socio-political and cultural perspectives, the environment was understood as conservation and the taming of the wilderness "before we abuse land because we regard it as a commodity belonging to us" (Leopold, 1994, p. 75).

Environmental policy in the UK was formed with the creation of a Department of Environment (DoE) (1970s). The first mention of environmentalism in the Queen's Speech took place in 1971 (McCormick, 1991) and the new government DoE was established. A White paper, *Reorganising of Central Government* (November 1970), was an early indication that governments were looking to administrate the population as an economic and social area of policy development. The creation of the DoE was not just part of a reorganizing structure by central government; more significantly, it was a reaction to increasing public concern around environmentalism. The new department symbolized the "advent of a new political awareness which recognised the socio-economic significance of environmental concerns" (Robinson, 1992, p. 11).

Despite its name, DoE's role focused less on environmental matters (such as conservation, air pollution, clean water supplies etc.) and more on overseeing infrastructure projects such as town planning, housing, inner-city issues, sport, recreation and royal parks (McCormick, 1991). The DoE's role was to administrate and organize societal structures, and management of the natural environment designed to accommodate population and urban growth for the benefit of humans.

Furthermore, governments took a more reformist, managerial approach to the environment and conservation, with a focus on green consumption (c.f. Elkington and Hailes, 1988). Managerialism means that power lies in the organization, and not with individuals. Organizations and corporations have enormous power in relation to the state, enhanced by creating their own lexicon. Power is not only in financial and regulatory terms, but also in terms of knowledge and expertise, and adeptness at language—a managerial language—which permeates society as well as state agencies, schools and universities. Political blogger, John Quiggin notes that

> The rise of managerialism has gone hand in hand with that of the radical program of market-oriented reforms variously referred to as Thatcherism, economic rationalism and neoliberalism. (Despite very different histories, all these terms are now generally used in a pejorative sense). Managerialism may appear inconsistent with traditional free-market thinking in which the ideal form of organisation is that of

> competitive markets supplied by small firms, in which the manager is also the owner. However, managerialism is entirely consistent with the dominant strand in the neoliberal approach to public policy, which takes the corporation, rather than the small owner-managed firm, as the model for all forms of economic and social organisation. (Quiggin,[2] 2003)

In other words, there's no difference between managing a soft drink and managing the environment when it comes to green economy. Green capitalism provides the neoliberal approach to public policy that becomes the norm for economic and social organization. Consumption is driven by the political centre right and right wing groups who take a lassie faire approach to the environment. The left, up until the birth of New Labour and the Blair government (1997) tended to take a more middle line position that matched that of professional ENGOs.

It is worth noting that green economics weren't only the preserve of neoliberal politicians. Professional environmental non-government organizations such as Friends of the Earth (FoE) and Greenpeace also took an economic approach by targeting "governments, business and consumers" (Price, Saunders, and Olcese, 2014, p. 170). Putting business at the forefront of their ideology is for some less a protest movement and more a protest business (Jordan and Maloney, 1997, p. 70) based on the idea that "contributors to Greenpeace are not formal members, but simply financial sponsors. Large organizations, such as Greenpeace and FoE, are best seen as having supporters rather than members" (Jordan and Maloney, 1997, p. 70). Thus, governance from combining managerialism and governance in managing the environmental also meant people became easier to manage and regulate—what Foucault terms biopower.

Biopower is "a mode of power that takes life as its object, while biopolitics refers to the techniques and rationalities of power mobilized in pursuit of the security, growth, and development of individual and collective life" (Foucault, 1978, 2003 in Grove, 2014, p. 22). From a governance position, biopolitics is the joining of biology, the body and politics, to generate techniques that discipline and persuade behavioural patterns of the general public in a way that governments feel is the best for society as a whole. One way that biopower and biopolitics have transpired since the late 1990s, is through a discourse of national security on the individual. The result is a new form of "biopower-security- emerges here to govern life through the interplay of freedom and danger" (Grove, 2014, p. 25).Individuals that chose not to conduct themselves to what's perceived as the benefit of wider society (conduct of conduct) are re-imagined as outside normative behaviour and therefore under biopower of security, it becomes justifiable for governments to coerce the body to develop mechanism and legislation. Biopower leads to power being no longer reducible to dealing with legal subjects but with "living beings, and the mastery it would be able to exercise over them, would have to be applied at the level of life itself" (Foucault, 1978, p. 143).

Government does not replace biopower, but forms a triangular power complex of sovereign–discipline–government or governmentality (Foucault in McNay, 1994, p. 1). Discipline is the triangulation of sovereign-discipline-power that emerges when individuals challenge the notion of conduct-of-conduct. Discipline can be suggestible through legislative or behavioral change. Such triangulation means that the truth of 'individuals' is no longer linked to the position they occupy in the universal order of things, but the individual is constructed around a "normalising notion of inner responsibility requiring an endless and thorough examination of the depths of their souls" (McNay, 1994, p. 28). Sovereign is "an anti-normative power, the power to appropriation" (Grove, 2014, p. 24). Sovereignty is the power to "make die and let live" (Foucault, 1977). Discipline is the way to exercise the power over the population, and government is not the "institutionalized organisation that hold power, but refers instead to the practices that attempt to modify the thoughts and cautions of others" (Grove, 2014, p. 23). Governments persuade the public through a very individualistic model which encourages each person to conduct themselves their family, their community, their employees in a way that's deemed appropriate for the betterment of society, meaning physical disciplinary measures are unnecessary.

Biopower provides practices and government policies on individual that encourages certain behavioural patterns, of greater benefit in the government's view for the wider society. Biopower is the "disciplinary techniques that operate on individual bodies, and governmental techniques that operate on the population. Together, discipline and government create particular forms of order by bringing individual capacities and mobilizations in line with normative visions of proper ways of being" (Grove, 2014, p. 24), for example, doorstop recycling or recycling supermarket bags, and ecolabels. We are encouraged to buy green through ecolabels to harness market forces by encouraging consumers to make a more informed purchasing decision. Ecolabels are regarded as un-intrusive policy instruments, with Germany the first to employ a national eco-label scheme

Biopolitics provides governmental the means to rationalize problems "governmental practices by the phenomena characteristic of a group of living human beings constituted as a population. (Foucault in Rabinow, 1997, p. 73). Therefore, Behavioural patterns can be nurtured through economics, so as Foucault says (1978, p. 41)

> without the controlled insertion of bodies into the machinery of production and the adjustment of the phenomena of population into economics process (…) it had to have methods of power capable of optimising forces, aptitudes, and life in general without at the same time making them more difficult to govern.

Biopolitics techniques and rationalities translate in environmental governance as positioning of activists as outside normative behaviour. Biopower by taking the life as its object and through different legislative techniques works as a form of power on entire populations. Biopower is a useful concept to show how environmentalism has increasingly become more about administrating life than using scientific solutions to climate change.

Case Studies: Economics and the Environment—A Stern Warning

The UK, USA and Australia have chosen to govern the environment through a green economic lens, that puts responsibility for the environment onto the individual. On this basis individuals would govern or conduct their behaviour in a way that reflects the priorities and rationalities defined by sustainability indicators, such as emissions trading schemes, or take upon eco-friendly practices such as door stop recycling or supermarket recycling schema, all of which require consumers to buy their way to environmental security.

Global institutions also put economics and environmentalism together in seeking solutions to the climate debate. The Club of Rome (1972), Brundtland Report in 1987, the Montreal Protocol on Substances that Deplete the Ozone Layer in 1989, Agenda 21 (1992) the Kyoto Protocol (1997 and into effect in 2006), became an affirmation of a discourse that "rests on assumptions, judgments, and contentions that provide a basic term for analysis, debates, arguments, and disagreements, in the environmental area" (Dryzek, 1997, p. 8). These initiatives to global environmental policies and practices each of these policy cycles were founded on a "truth" that the global climate is changing because of human impact. So climate change aggregated by human industrialization became a discourse based on a scientific truth, reinforced by government and institutional policies.

In the United States, President George W Bush signed into American law the Energy Policy Act (2005) which amongst other policies meant an increase in clean coal and biofuels. In the USA, Clinton and Gore's "Global Marshall Plan" for the environment was meant to stabilize the "world population, by developing environmentally appropriate technologies, a change in economic 'rules of the road' of how environmental impact is measured and a call for new international agreements" (Gore, 1992, pp. 297–301).

Australian Labor Prime Minister, Kevin Rudd ratified the Kyoto Protocol (2006), after his predecessors having initially agreed to the protocol, later refusing to ratify on the grounds of economic harm to the country. Prior to Rudd's signing, In Australia, Prime Minister, John Howard creates the Australian Greenhouse Office (AGO) (1998). A world first, the AGO was "the key Commonwealth agency on greenhouse matters, responsible for both the coordination of domestic climate change policy and for managing the delivery of major new and existing Commonwealth greenhouse programmes"[3]. Climate change solutions were being coalesced into domestic economic development plans, as politicians turned to the markets to find global warming solutions.

Green Governmentality: Vote Blue, Go Green

A decade of environmental policies and growth as public calls for more information peaked in UK at the 2006 local election (May). These elections were unique as they made environmentalism a key issue, as all three main parties began chasing the green vote in marginal seats (Jacobs, 2010). Media reports of Conservative party leader, David Cameron sledging on Norwegian Fjords and cycling to his parliamentary office at Westminster Palace, was aiming to reinforce a conduct-of-conduct approach—lead by example. Environmental stories were dominant in the news media, as A whale swam up the Thames river (January), green entrepreneur Al Gore's film An Inconvenient Truth (Gore, Guggenheim, David, Bender, Burns and Skoll, 2006) was a global success (September). Consequently, celebrities and politicians were all vying over environmental discourse through public spectacles and events.

Former American Vice President Al Gore's documentary *An Inconvenient Truth* (2007) was soon followed by the global music extravaganza Live Eight to coincide with the Group of Eight (G8) meeting in Scotland, UK. A meeting of the leaders and financiers from the eighth wealthiest countries in the world. Green businessman Gore's film inspired rock band Razorlight's singer Johnny Borrell to play at a Live 8, applauding Gore's film for raising awareness: Razorlight headlined a pop concert that was the culmination of a march through central London to Trafalgar Square. The concert, organized by professional green groups, Greenpeace, Stop Climate Chaos coalition, and the Friends of the Earth (including their Big Ask campaign of May 2005) was a culmination of a sustained campaign calling for a UK Climate Change Bill made law through new UK legislation.

Conservative Party's then Director of Strategy, Steve Hilton, once described by the *Daily Mirror* newspaper as "the puppet master who has attempted to transform pin-striped Old Etonian Cameron into dressed down 'Dave', man of the people" (Brough, 2010), was instrumental in reinventing the Conservative Party within environmental discourse. It was a political move that had succeeded a decade earlier with Prime Minister Margaret Thatcher's Green Speech (1988) on the depleting ozone layer at the Royal Academy of Science. The reincarnation of the Tories as environmental conscious party was led by former Friends of the Earth staffer, Steve Hilton.

Hilton is credited with reinventing the UK Conservative Party under the slogan "Vote Blue, Go Green". The tag line was alongside a new logo of a tree, replacing the traditional Conservative party oak tree logo. The oak tree logo stood for stability and longevity of the party; whereas Vote Blue, Go Green was part an outward political shift to embrace the climate change debate within party politics and the media. The *Daily Mirror* newspaper saw the new logo and campaign as a way to demonstrate under the slogan Vote Blue, Go Green, Cameron insisted his party had the best environmental policies. He said 'if you want more recycling, greener parks and public spaces, cleaner streets and better transport, vote Tory[4]' (anon, 2012, 19-05-2006). The cleaner streets, public space and greener parks a revisiting of Thatcher's Tidy Britain litter campaign. The *Daily Mirror*[5] covers the cycling Cameron story with the headline, "Con Yer Bike Cameron'—'green' Tory Dave Cameron cycles to work and attacks mothers on school run … then have his driver bring his shoes to office in a gas-guzzler" (Maclean and Robert, 2006, p. 11). The *Daily Mirror* accuse Cameron of fakery and a media stunt as "The Tory leader, desperate to hijack the green vote' at next Thursdays local elections" (ibid). Cameron like Thatcher was using the electorate to convince the individuals voting that "tidy", clean environmental policies would be of benefit to all, and physically using his body (cycling) to demonstrate his message.

The May 2006 local elections became a platform to launch a new Conservative party brand. The previous logo, a freedom torch introduced by Thatcher in 1977, updated in 2004 by then leader Howard became an oak tree (Browne, 2006). The choice of an oak tree stood for "solidarity, tradition, friendliness towards the environment and Britishness" (Browne, 2006). A media campaign saw Cameron cycling to work and visiting Norwegian Fjords. The new logo was launched as a "rebranding of the party, aimed at promoting a more modern, softer, environmentally friendly image", reports *The Times* journalists, Anthony Browne (2006). *The Times* highlights how "the decision by David Cameron to cycle to work each day symbolized the reinvention of the Tories as the blue party that thought green", but it was revealed as Cameron cycles to Parliament and a press pack, he's followed by "Mr Cameron's chauffeur to collect a package. Two minutes later he set off for Westminster bearing a pair of polished shoes, a brief case and a crisp new shirt for his lycra clad boss" (Browne, 2006, p. 27).

In reality, Cameron was mirroring his predecessor Margaret Thatcher and her earlier Tidy Britain campaign as an example of enforcing the biopolitics of a conduct of conduct. Thatcher linked environmentalism to family values to encourage the public to have "conservation, heritage and English values" (Watt, 1999, p. 86). Thatcher's Tidy Britain Campaign (1988), was a way to present a "conduct of conduct" by linking litter as a metaphor for national pride and stability. Thatcher says, "Litter is everyone's problem in which everyone must contribute to a solution. It is like good manners; in my opinion … I want to see pressure coming from below, from street level, from village level, from individuals, and to see that pressure converted into action. If we could do this, it would make more laws and more penalties unnecessary" (Thatcher Press Conference, 22 March 1988).

The same principle has been applied by Consecutive UK governments (vis-à-vis Conservative and New Labour), international organizations (such as the United Nations) and multinational corporations understood that if you can coax and persuade individuals to consume eco-friendly products on the understanding it benefits the wider society would result in higher profits. British Prime Ministers from Margaret Thatcher, to Tony Blair, Gordon Brown, and David Cameron came to realize that economics solutions to climate change can both open up new markets and increase consumption in eco products.

Thatcher had always seen the environment as a means of economic growth and business opportunities. Although Thatcher firmly believed Kyoto was an "anti-growth, anti-capitalist, anti-American project" (Thatcher, 2003, p. 453), it did generate new markets that acted as technologies on the individual. Putting economics at the forefront of global warming solutions Thatcher first conflates climate change solutions with economics in her green speech (1988) to the Royal Academy of Science.

Following the discovery, by British scientists of a hole in the earth's ozone layer (1988) Thatcher set out plans to solving "global warming" through collaboration between businesses and science. Thatcher's speech had two aims, firstly to put Britain on the global stage for climate change solutions; Secondly, to introduce deregulation of energy markets and increase freedom in regulatory rules, including oil and financial industries. With greater market fluidity, meant, "boundaries between different parts of companies, between different companies, and between companies, governments and NGOs, are broken down as actors seek new ways of solving problems" (Newell and Paterson, 2010, p. 22). Thatcher's "markets, monetarism and authoritative" form of government (Luke, 1999, p. 159) looked towards enterprise to find solutions to global warming and the depleting ozone layer by encouraging business leaders to work with scientists (Newlands, 2012a, 2012b). The speech says,

> helping] our academics to spot commercial an application … Industry is becoming more scientific-minded … scientists more industry-minded. Both have a responsibility to recognize the practical value of the ideas, which are being developed. (Thatcher Press Conference, 22 March 1988)

Carvalho (2007) suggests Thatcher's co-opting of environmentalism was in reality an attempt to push forward nuclear power and the privatization of markets agendas. Lester (2010) supports Carvalho's claim that Thatcher's "appropriation of the risk of climate change" helped to "support a case for nuclear energy over coal and thus weaken the coal industry" (2010, p. 67). Thatcher's proposal for solutions to come from business and science meant devolving responsibility for environmental problems from state to business and down to the consumer. Thatcher's speech centred on "addressing the absence of leadership on the international environment" (McCormick, 1991, p. 65), suggesting that British business might take leadership over climate change solutions. Thatcher's speech had another objective-to place the UK at the forefront of climate science. The speech stresses a need for international co-operation in tackling climate change and global warming, whilst reminding the world it was "British scientists [who] had discovered a thinning of the ozone layer over the Antarctic" (McCormick, 1991, p. 104). Thatcher's speech meant greater media coverage to climate change solutions, but also replaces scientists with political actors increasingly seeking to influence the agenda (Carvalho and Burgess, 2005). Political parties, like climate skeptics began to realize that many government policies green tinge (Watt, 1999) was enough to give the appearance of being environmentally aware.

Thatcher was making the most of her science background to give authority to the campaign, but some journalists and experts were critical of this apparent change of heart. Indeed, until the green speech, it was clear that Thatcher's early administration had greater enthusiasm for war than climate change. In discussions on the Falklands War in 1982, Thatcher defined climate change solutions as "humdrum", in comparison to dealing with conflict—it was more "exciting to have a real crisis on your hands" (Robinson, 1992, p. 177). Former Director of Friends of the Earth, the Hon. Sir Jonathon Porritt, 2nd Baronet, CBE says, "making statements on the international environment ... involved far fewer policy commitments than statements on the domestic environment" (McCormick, 1991, p. 66). The "green" speech served a useful purpose in pushing forward Thatcher's own "style of how the nation should be run" (Robinson, 1992, p. 176), and, like war, gave a sense of immediacy to an open acceptance of the scientific arguments. The Green speech was also tapping into and leading the way for emerging global political discourse. In positioning the UK within an emerging environmental discourse, Thatcher set the UK as environmental, against a "backdrop of drought in the United States and unusual weather patterns" (Grubb, 1998, p. 1). Nearly ten years later, New Labour's commissioning of the Stern Review to assess economic impact of climate change would put their neoliberal policies at the forefront of the debate.(2006a, 2006b, 2006c).

The Stern Review (2006a, 2006b, 2006c) set out to establish the economic impacts of ignoring climate change. Lord Stern's report created an economic discourse for markets to focus on climate issues. The findings showed that each tonne of CO_2 that the UK emits causes at least US$85 trillion worth of damage to the nation's economy (Stern, 2006a). The report recommended that a cut in emissions would limit the economic impact of climate change to approximately 1% of global GDP. Stern proposed to offset the cost by charging individuals more for carbon-intensive goods and by investment in new technology, "a techno-fix approach" (Stern, 2006a). A low-carbon economy would ultimately be less costly than the current high-carbon economy, which spoke to the dominant discourse, generated political agents and supported the climate change argument. Shifting the world onto a low-carbon path could eventually benefit the UK economy by $2.5 trillion a year and, by 2050, markets for low-carbon technologies could be worth at least $500bn a year. However, subsidizing or reducing aviation emissions was deemed too costly (Stern, 2006a).

Lord Stern's review also reinforced the notion of biopolitics, because it shifted the emphasis onto the individual through neoliberal or advanced liberal government. Through a series of separate micro-systems, individuals are given targets or 6% reduction in homes and small businesses. The targets aimed at promoting the increased use of renewable resources in homes and small businesses, and a small-scale renewable target of 6% in all homes by 2020. Where possible, dwellings in UK should have proper insulation by 2015, and each household must install smart-meters to measure energy consumption, an idea earlier mooted by Cameron. Accelerating production on electric-powered vehicles, hybrid vehicles, were part of a low-carbon industrial strategy, along with the creation of a new financial institution (a green bank). The UK government sold the Green Bank to a private investment firm in August 2017.

The Environment as a New Mode of European Managerialism

The Climate Change Act and Lord Stern's report cements economics to energy, development of new markets, and makes capitalism the solution to climate change. Yet, none of the markets are regulated (Newlands, 2012a) and like the CoP agreements are taken on trust.

By responding to changes in domestic and global governance in seeking ways to follow the Conservatives in linking economics to the environment, New labour sought ways to go "green". New Labour executives were concerned the Green Party would sway voters in marginal seats. Partly fearing a public backlash for not being green and worries over marginal seats, Chancellor, Gordon Brown announces the UK's first Climate Change Act (2007–2008). Pressure from NGOs and the Climate Coalition and encouraged by New Labour's former Environment Secretaries, David and Ed Miliband and Special Advisor Michael Jacobs (c.f Jacobs, 2010), New Labour under Brown, took steps to reduce carbon emissions with a Climate Change Act (2007–2008).

The Climate Change Act's opening gambit declares, "The threat from climate change is perhaps the greatest challenge facing our world"[8] and, echoing Gore's earlier rhetoric. The new legislation was daring and radical by the New Labour Party, which set out a number of key objectives and targets. The Act recommends creating a carbon budget system to reduce carbon emission by 2050. Carbon reduction targets were established by an independent committee on climate change. The committee's aims were a 34% reduction in greenhouse gas emissions in all areas by devising a low carbon transition plan. The Act also gave government the powers to establish trading schemes for limiting greenhouse gas, but was disbanded by 2016.

The UK continued to lead the new green economy, pioneering emissions trading instruments. Trading in emissions on a global scale was an idea first developed in the UK by Michael Grubb (1998), and developed by the City of London, and the UK government. Grubb drew on an earlier North American plan to address the problem of acid rain and identified that any decline in emissions needed a multilateral approach between countries and global companies rather than individuals, states or regions. Grubb proposed a new emissions trading scheme (ETS) or "cap and trade"—"the buying and selling of pollutant entitlements" (Newell and Paterson, 2010, p. 96)—that focused on a division of emissions between countries, through a system of "contraction and convergence" first developed by London company the Global Commons Institute (GCI). The mechanisms for 'contraction and convergence' came from the London markets. The contraction and convergence model argues that if one country produces fewer emissions than others do they could increase emissions allowances for developing countries to create an equilibrium in global emissions through trade. The scheme was agreed and formalized through the Kyoto Protocol, along with a series of mechanisms to lessen carbon emissions: 1) Clean Development Mechanisms;[6] 2) Emissions Trading Schemes; and 3) Joint Implementation.[7] Of the three approaches, the EU took the lead in emissions trading schemes and new environmental policy instruments. Along with the Netherlands and the EU,[9] the UK created the European Union Emissions Trading Scheme (EU ETS). As a result, enterprises increasingly found new ways to incorporate environmentalism into business plans. The governments still held a role in legislating, with carbon markets and emissions trading becoming rapidly growing areas in market investment. Global Environmental Governance policies and developing market-led initiatives supported, in turn, by UN initiatives such as Agenda 21 (2003), the Brundtland Commission

(1989) and the Kyoto Protocol (1992), meant environmentalism became "embedded" in global industrial and institutional practices. At the forefront of this development was the City of London. London is home to many major carbon-trading organizations, creating billions of dollars of trade a year (Newlands, 2012). However, the systems were open to manipulation and soon there were accusations of Greenwashing.

Greenwashing and the Corporation

Accusations of corruption emerge around companies who after purchasing permits shelve plans intended for the permits, leaving surplus permits to sell on the markets. For example, steel manufacturer Corus emissions permits for development at their Teesside plant, shelved their plans, leaving them able to sell the unused permits for an estimated profit of £250 million (Clover, 2010). Moreover, with the global recession came a decrease in demand for energy. Companies and institutions that couldn't raise capital through bank loans were able to generate funding by "selling their allowance-gambling that they would be able to buy [...] back when customers returned" (Clover, 2010). Moreover, once the third phase of ETS came into effect in 2013, organizations with remaining permits were able to "carry over 1.8 billion permits ... obviating the need to buy any new credits before 2016" (Schiller, 2011). This profiteering on permits shows how a reduction in energy consumption means a decrease in emissions freeing up more permits to sell on the carbon markets. Moreover, the reliance on networks and trade in non-tangible products leaves the systems open to abuse. In February 2011, the ETS system and nine participating countries were subject to a cyber-attack. A series of phishing emails convinced many companies to sell their emissions allowance, and the Financial Times Deutschland reported that one firm had lost €1.5 million as a result (Philips, 2010). However, these examples of did little to deter the growth of emissions trading, thus avoiding detriment to the market.

The business case for a Climate Change Act was led by the commercial energy sector needing fresh and new forms of investment. In creating a strong business agenda for climate change, green governmentality generated new markets and also new investment opportunities for corporations. The creation of emissions—trading markets opened up new avenues of investment for large corporations, whilst fulfilling their corporate mandates on social responsibility. There was also an element of moral—suasion (a term that has similar meaning to persuasion) in the rhetoric. The adoption of a green agenda by multinational corporations meant they would appear to be pushing a green agenda, whilst expanding new markets.

Lord Stern's report, the Climate Change Act, the Oak Tree and the Kyoto Protocol all seek solutions to the climate debate through bio-politics. Whilst the concept and practices of biopower are useful in identifying how power relations work within environmental discourse, they are also useful in unpacking the shift in power relations between activists and mainstream politics.

Green Governmentality and an Uncertain Future for EcoActivists

As we continue to live in a society founded on high-carbon production. Toffler (1984) would argue we are living in a "third wave" era of a post-industrial, technologically led society, with rapid growth reliant on a high carbon economy. Since the Cold War there has been a "more accelerated industrial development" (Finger, 2008, p. 44) creating a society based on mass consumption, production, media and the growth of technological advances. This "Third Wave" (Toffler, 1984) post-industrial, technologically driven society, like its predecessors, is dependent on a high-carbon economy to grow. The increase in consumption means that "increases in energy use, the transportation of goods, the heating of houses, the powering of industries, and so on" would all call into question our use and abuse of fossil fuels (Newell and Paterson, 2010, p. 14).

The early environmental movements, such as Greenpeace, Reclaim the Streets, Critical mass, CND, Friends of the Earth, and the radical movements continued to focus on environmental social justice; whereas governments and global institutions saw environmentalism through the lens of capital and governance. This results is that multinational companies and governments' interpretations of environmentalism has shifted focus away from relationships between people and the natural environment and towards a discourse that centres on the administration and regulation of life (biopower).

By administrating life and regulating the individual through government departments, and globally through Multilateral environmental agreements (MEAs), environmental protocols (such as Basel, Montreal and Kyoto) and global institutions and organizations such as the IPCC, UNEC) and the United Nations Framework Convention on Climate Change (UNFCCCP) policies and governance develops to make it the individual responsibility for climate change.

The adoption of a green agenda by multinational corporations meant they would appear to be pushing a green agenda, whilst expanding new markets. At the same time there was need to protect big business from protests and activism, which would be economically damaging for the corporations.

Dryzek, Norgaard, and Schlosberg (2013) suggest that in light of climate change there is a need for a "radical reworking on the very nature of governance" (2013, p. 95). Dryzek, Norgaard and Schlosberg main premise on governance emerges from the idea that climate change as a discourse present us with a "mirror in which we can see the best and worst of human society", and that examining governance through the lens of climate change enables us to see how "human activity that increase shapes the global environmental in which we live" (2013, p. v). Therefore governance needs to understand in terms of global environmental change brought on by the science of climate change over political discourses. An Australian context where governance is understood more as political not global governance issue leads Dryzek, Norgaard and Schlosberg argue what the "Australian case really illustrates is the difficulty that political systems used to treating every issue as grist for a two-sided partisan electoral contest have when it comes to processing knowledge on an issue like climate change" (2013, p. 36).

Furthermore, Dryzek, Norgaard and Schlosberg (2013) propose three distinct areas that need looking at in order to rework governance, 1) national governments, 2) global governance and 3) networked governance. National governments react to climate change and governance in two ways, 1) a wait and see approach and 2) lack of prioritization when it comes to environmental conservation. National governments often wish to wait and see if other governments will take the lead on implementing new or amended environmental policies (2013, p. 107).

In addition the role of NGOs' social movements, an activists and the media play an important role in the public space, that "together these help constitute a larger 'public sphere' that is a source of both ideas about what to do, and pressure on governments actually to do it" (Dryzek, Norgaard and Schlosberg (2013, p. 101). However, although national governments will have to initiate policy at a local level, what happens with global environmental change will impact on national policies. Dryzek, Norgaard and Schlosberg note the "networks involving both public sector and corporate actors are also emerging to explore and promote renewable energy, geoengineering, and carbon capture and storage" (2013, p. 109). Dryzek, Norgaard and Schlosberg conclude that "reconstructing governance for a climate challenged society remains a major task" (2013, p. 110).

The result is that we get a series of rules and regulations—or governance—rule setting that defines what acceptable and unacceptable environmentalism is. The environment becomes a combination of political decision-making a, big business protectionism and as a consequence the increasing criminalization of environmental activism. Governance with an environmental flavour has had consequences on environmental activists' movements in the UK. The neoliberal ideology of late twentieth century Britain has become so intertwined that some environmental activism has become a criminal offence. The lengths police and politicians have gone to, to protect big business ranges from filming activists at protest, to undercover policing, such as the Mark Kennedy case (see Chapter Four), and the framing of activists as domestic extremists.

New Labour as well as pushing a green capitalist agenda, was simultaneously increasing surveillance and criminalizing protest. At the turn of the century, three years after Tony Blair's New Labour government came to power; there was an unprecedented increase in the undercover infiltration of political activists. There was no apparent rise in the threat from radical politics to explain the change in gear. But protest was no more or less threatening than it had been in previous years. The Special Demonstration Squad was still going strong. The unit created by Conrad Dixon (Lewis and Evans, 2013, p. 199)

> under the Labour government was no different. In the first decade of the 21st century, funding for policing domestic extremists doubled. The breadth of activity being targeted by police also expanded. Documents that chronicle the birth of the NPOIU reveal it initially intended to focus upon two groups. hard-core animal rights activists behind sometimes nasty campaigns against vivisectionists and a group defined as "environmental extremists", taken to be a loose reference to the anti-roads movement. (Lewis and Evans, 2013, p. 202)

Lord Stern's report, the Climate Change Act, the Oak Tree and the Kyoto Protocol all seek solutions to the climate debate through bio-politics. Whilst the concept and practices of biopower are useful in identifying how power relations work within environmental discourse, they are also useful in unpacking the shift in power relations between activists and mainstream politics.

The biopolitically inspired neoliberal ideology of late twentieth century has become so intertwined with eco-or green governmentality to the point that in the battle over who dominate environmental politics, activism has become a criminal offence. The lengths police and politicians have gone to protecting multinational corporations ranges from filming activists at protest, to undercover policing, such as the Mark Kennedy case, and the framing of activists as domestic extremists (see Chapter Four). The result is that a manifestation of truth emerges through the media and political decision making as to what has become acceptable and unacceptable environmentalism.

Conclusion

The purpose of this chapter has been to show how the state's application of neoliberal concepts to environmental discourse has excluded horizontal ecoActivism from environmental discourse. Over time, party politics, local and global institutions, large corporations and cross-party policies have used eco-power and eco-knowledge to create new markets in a discursive struggle between activist and governance. This chapter has illustrated how the UK government's co-opting of environmental discourse through the scientific data develops Luke's argument that power/knowledge in an environmental context translates into eco-power alongside eco-knowledge (1999), the chapter has explored how power relations can aid discursive shifts. As power is omnipresent, and it has argued that power is not about ownership or control but relations between two contrasting sides.

Through the discussion on biopower, this chapter has shown how the UK government began to persuade individuals to engage with environmental discourse through the concept of neoliberalism. In charting how neoliberalism became central to environmental discourse, this chapter has shown the increasing role of economics in relation to environmentalism. Moreover, shifts in power can be seen through technologies of governmentality that provide the mechanism to go green, whilst at the same time positioning activists as abnormal, delinquent and deviant. Once the domain of activists through the media, environmental discourse has changed from activism to advanced liberal government as the leading policy. Biopower allows the state to frame activists as delinquent and deviant reinforced through technologies of dominance, whilst encouraging individual to consume ethical eco-friendly lifestyles.

Moreover, the Climate Change Act (2007–2008), provides a platform for economic and market-led environmental discourse. The business case for a Climate Change Act existed due to the commercial energy sector needing fresh and new forms of investment. A business as usual approach to climate change solution prevails today. In creating a strong business agenda for climate change, green governmentality generated new markets and also new investment opportunities for corporations. The creation of emissions—trading markets opened up new avenues of investment for large corporations, whilst fulfilling their corporate mandates on social responsibility. The Act provided new markets, in the form of NEPIs, GEG and PPP, to reinforce economics as the dominant position of environmental discourse. Furthermore, this action is mirrored on a global scale, with the Kyoto Protocol's launching of other market-led solutions to climate change (such as EU ETS and carbon trading). This chapter has focused on plotting the historical process of green capitalism and its creation of new markets. The next chapter will look at the consequences that an increased economic role in environmental discourse has had on the environmental activists' movements. It will explore how the exclusion of environmental activists from environmental discourse that was discussed here, is repeatedly reiterated through technologies of dominance predominantly found in legislative and state apparatus. The next chapter looks at the other side of the debate (technologies of dominance) the disciplining of the biopolitics through state surveillance and the linking of ecoActivism with terrorism and the nomenclature of "ecoterrorism". The counter or flipside of eco-governance and the legitimization of criminalizing eco-activism "recasts" individuals taking part in protest or action as domestic extremists'.

Notes

1. See http://unfccc.int/bodies/body/6383.php
2. See http://johnquiggin.com/2003/07/02/word-for-wednesday-managerialism-definition/
3. http://parlinfo.aph.gov.au/parlInfo/search/display/display.w3p;query=Id%3A%22media%2Fpressrel%2F7V505%22
4. http://www.mirror.co.uk/news/uk-news/cameron-changes-to-green-621439
5. http://www.mirror.co.uk/news/uk-news/exclusive-con-yer-bike-cameron-622779
6. CDM project activity might involve, for example, a rural electrification project using solar panels or the installation of more energy-efficient boilers. The mechanism stimulates sustainable development and emission reductions, while giving industrialized countries some flexibility in how they meet their emission reduction or limitation targets.

7. The mechanism known as 'joint implementation', defined in Article 6 of the Kyoto Protocol, allows a country with an emission reduction or limitation commitment under the Kyoto Protocol (Annex B Party) to earn emission reduction units (ERUs) from an emission reduction or emission removal project in another Annex B Party, each equivalent to one tonne of CO2, which can be counted towards meeting its Kyoto target. Joint implementation offers Parties a flexible and cost-efficient means of fulfilling a part of their Kyoto commitments, while the host Party benefits from foreign investment and technology transfer. (http://unfccc.int/kyoto_protocol/mechanisms/joint_implementation/items/1674.php) accessed March 2001
8. See Great Britain. *The Climate Change Act (2007–8)*, London: http://www.publications.parliament.uk/pa/ld200708/ldhansrd/text/81126-0006.htm#08112648000004 (accessed June 2010) for further details
9. They were followed by Austria, Finland, France and Ireland (Eire).

References

Bevir, M., and Rhodes, R. (2003). *Interpreting British Governance*, London: Routledge.

Brough, G. (2010). "David Cameron's Chief Spin Doctor Steve Hilton: The Shameful Secrets of Cyclopath". Available at http://www.mirror.co.uk/news/top-stories/2010/01/08/david-cameron-s-chief-spin-doctor-steve-hilton-the-shameful-secrets-of-cyclopath-115875–21951336/ (accessed January 2011).

Browne, A. (2006). "Tories Go for Oak in an Attempt to Soften Their Image", *The Times* (London), September 15, p. 2.

Buckingham, S., and Turner, M. (2008). *Understanding Environmental Issues*, London: Sage.

Carvalho, A. (2007). "Ideological Cultures and Media Discourses on Scientific Knowledge", *Public Understanding of Science*, 16(2), pp. 223–243.

Carvalho, A., and Burgess, J. (2005). "Cultural Circuits of Climate Change in UK Broadsheet Newspapers, 1985–2003", *Risk Analysis*, 25(6), pp. 1457–1469.

Clover, C. (2010). "Buying Off Big Polluters Looks Bad But It Works", *The Sunday Times Newspaper*, National Edition, 28 February, p. 21.

Cox, R. (2010). *Environmental Communication and the Public Sphere*, Thousand Oaks, CA: Sage.

Darier, E. (1999). *Discourses of the Environment*, Oxford: Blackwell.

Dryzek, J. (1997). *The Politics of the Earth: Environmental Discourses*, Oxford: Oxford University Press.

Dryzek, J., Norgaard, Richard B., author, & Schlosberg, David, author. (2013). *Climate-challenged society / by John S. Dryzek, Richard B. Norgaard, and David Schlosberg*. (1st ed.).

Elkington, J., and Hailes, J. (1988). *The Green Consumer Guide: From Shampoo to Champagne: High-Street Shopping for a Better Environment*, London: Gollancz.

Finer, S. (1970). *Comparative Government*, Harmondsworth: Penguin.

Finger, M. (2008). "Which Governance for Sustainable Development? An Organizational and Institutional Perspective". In: J. Park, K. Conca, and M. Finger, *The Crisis of Global Environmental Governance: Towards a New Political Economy of Sustainability*, New York, Taylor Francis, pp. 34–

Foucault, M. (1977). *Discipline and Punish: The Birth of the Prison*, London: Penguin.

Foucault, M. (1978). *The Will to Knowledge: The History of Sexuality:1*, translated by R. Hurley, London: Penguin.

Gore, A. (1992). *Earth in the Balance: Forging a New Common Purpose*, London: Earthscan Publications.

Gore, A., Guggenheim, D., David, L., Bender, L., Burns, S. Z., Skoll, J ... Paramount Classics (Firm) (Directors). (2007). [Video/DVD] Australia: Paramount Home Entertainment (Australasia).

Great Britain. *The Climate Change Act (2007–8)*, London. Available at http://www.publications.parliament.uk/pa/ld200708/ldhansrd/text/81126-0006.htm#08112648000004

Grove, K. (2014). "Biopolitics", in C. Death (ed.), *Critical Environmental Politics*, London, New York, Routledge: Taylor & Francis Group.

Grubb, M. (1998). *Corrupting the Climate? Economic Theory and the Politics of the Kyoto.* The Royal Institute of International Affairs; Energy and Environmental Programme, Chatham House. Available at http://www.econ.cam.ac.uk/rstaff/grubb/publications/CPWP34.pdf (accessed December 2010).

Jacobs, M. (1991). *The Green Economy, Environment, Sustainable Development and the Politics of the Future*, London: Pluto Press.

Jacobs, M. (2010). Personal communication. Meeting of the ECPR Environmental Politics and Policy Summer School, Keele University, June.

Jordan, G., and Maloney, W. (1997). *The Protest Business? Mobilising Campaign Groups*, Manchester: University of Manchester Press.

Jordan, A., Wurzel, R., and Zito, A. (2003). "'New' Environmental Policy Instruments: Patterns and Pathways of Change", *Environmental Politics*, Special Issue, 12(1), pp. 1–25.

Kelly, G. E. M. (2009). *The Political Philosophy of Michal Foucault*, London: Routledge.

Leopold, A. (1994). "The Land Ethic", in D. Wall (ed.), *Green History: A Reader in Environmental Literature, Philosophy and Politics*, London and New York: Routledge.

Lester, L. (2010). *Media and Environment: Conflict, Politics and the News*, Cambridge: Polity.

Lewis, P., and Evans, P. (2013). *Undercover: The True Story of Britain's Secret Police*, London: Faber and Faber.

Luke, T. (1999). "Environmentality as Green Governmentality", in E. Darier (ed.), *Discourses of the Environment*, Oxford: Blackwell, pp. 121–151.

Maclean, S., and Robert, B. (2006). 'Con Yer Bike Cameron', *Daily Mirror*. Available at http://www.mirror.co.uk/news/uk-news/exclusive-con-yer-bike-cameron-622779

McCormick, J. (1991). *British Politics and the Environment*, London: Earthscan Publications.

McNay, L. (1994). *Foucault. A Critical Introduction*, New York, NY: Polity Press.

Newell, P., and Paterson, M. (2010). *Climate Capitalism: Global Warming and the Transformation of the Global Economy*, Cambridge: Cambridge University Press.

Newlands, M. (2012a). 'Carbon Trading in the City of Undoing', in G. Poynter, I. MacRury, and A. Calcutt (eds.), *London After Recession: A Fictitious Capital?*, London: Ashgate Publishers, pp. 199–213.

Newlands, M. (2012b). 'Reclaiming the Media: Technology, Tactics and Subversion', in E. Cudworth, P. Senker, and K. Walker (eds.), *New Horizons: Technology, Society and Inequalities*, London: Peter Lang.

Oels, A. (2005). "Climate Change Governable: From Biopower to Advanced Liberal Government", *Journal of Environmental Policy and Planning*, 7(3), pp. 185–207.

Philips,L.(2010) Cyber-scam artists disrupt emissions trading across EU, at Carbonwatch, available at http://www.carbontradewatch.org/in-the-media/cyber-scam-artists-disrupt-emissions-trading-across-eu.html

Pierre, J., and Peters, B. G. (2000). *Governance, Politics and the State*, Basingstoke: Macmillan.

Price, Saunders, and Olcese (2014). "Movements", in C. Death (ed.), *Critical Environmental Politics*, London, New York, NY, Routledge: Taylor & Francis Group.

Rabinow, P. (ed.). (1997). *Michel Foucault Ethics, Essential Works of Foucault 1954–1984*, Vol. 1, London: Penguin.

Robinson, M. (1992). *The Greening of British Party Politics*, New York, NY, USA: Distributed exclusively in the USA and Canada by St. Martin's Press, Manchester: Manchester University Press.

Rhodes, R. A. W. (1996). "The New Governance: Governing without Government", *Political Studies*, 44(4), pp. 652–667.

Schiller, B. (2011). "Is It Time to Overhaul Europe's Carbon Trading Scheme?", *Guardian.co.uk*, April 28. Available at https://www.theguardian.com/environment/2011/apr/28/overhaul-europe-carbon-trading-scheme

Schout, A., and Jordan, A. (2005). "Coordinate European Governance: Self Organised or Centrally Steered?", *Public Administration*, 83(1), pp. 201–220.

Solomon, S., Qin, D., Manning, M., Chen, Z., Marquis, M., Averyt, K. B., Tignor, M., and Miller, H. L. (eds.). (2007). *Contribution of Working Group I to the Fourth Assessment Report of the Intergovernmental Panel on Climate Change*, Cambridge, UK, and New York, NY: Cambridge University Press.

Stern, N. (2006a). *The Economics of Climate Change: The Stern Review*, Cambridge: Cambridge University Press.

Stern, N. (2006b). *Review on the* Economics of Climate Change, HM Treasury, London: HMSO.

Stern, N. (2006c). *Review on the* Economics of Climate Change: Executive Summary, HM Treasury, London, HMSO.

Stoker, G. (1998). "Governance as Theory", *International Social Science Journal*, Vol 50, Issue 155, pp. 17–28.

Thatcher, M. (1988). Margaret Thatcher's public speech to the Royal Society, Fishmongers Hall, City of London, 27 September 1988. Available at http://www.margaretthatcher.org/speeches/displaydocument.asp?docid=107346 (accessed 30 September 2009).

Thatcher, M. (2003). *Statecraft: Strategies for a Changing World*, London: Harper Collins.

Toffler, A. (1984). *The Third Wave*, New York, NY: Bantam Books.

Watt, D. (1999). *The Environment and British Politics*, London: Hodder & Stoughton.

· 4 ·

RECASTING ENVIRONMENTAL ACTIVISM AS CRIMINAL DISSENT

Soft Power, Political Policing and the Media

> We assume at every meeting there are at least one journalist and one special branch officer.
>
> —Jan (interview, 2015)

> On at least two occasions, there were as many corporate spies at meetings of the small group as genuine activists. The number of real activists in the meetings had by then dwindled to between five and ten, and sometimes fewer. The spies had an incentive to put more effort into building their credibility in the group and so often attended more meetings than bona fide campaigners.
>
> —Lewis and Evans (2013, p. 71)

October 2010 and a notice appears on the alternative news website, Indynews, Nottingham (IndynewsNotts), it reads as follows;

> Mark "Stone" has been an undercover police officer from 2000 to at least the end of 2009. We are unsure whether he is still a serving police officer or not. His real name is Mark Kennedy. Investigations into this identity revealed evidence that he has been a police officer, and a face-to-face confession has confirmed this. Mark claims that he left the police force in late 2009 and that before becoming an undercover officer he was a Metropolitan police constable. Please pass this information on to anyone who may have been in contact with Mark in the last decade, both in the UK and abroad. (IndynewsNotts)[1]

The National Public Order Intelligence Unit (NPOIU) tasked PC Kennedy and others to gather intelligence on various environmental activist collectives across Europe. Kennedy's exposure confirmed what many activists had suspected, that governments were using political policing (undercover police) to gather intelligence data on civil disobedience and direct action collectives. Kennedy was just one of a historical chain of undercover police stretching back to the 1960s. Since the 1990s UK Police force have put a lot of focus on the environmental movements, and as this chapter will show, there has been unproven connections between ecoActivism and terrorism and the nomenclature of "ecoterrorism" (Arnold, 1987).

This chapter looks at the conflating of various movements and concepts through the lens of disciplinary powers (technique of dominance in Foucauldian terms, 1977). This chapter will first look at the rise of anti-protest laws from Australia and New Zealand, USA, and UK before looking at two case studies of when control over the individual or a group has failed to prevent protests from going ahead. This chapter identifies and examines discursive moments within the history of political policing to chart the trajectory from undercover policing in the 1960 to linking ecoActivism with terrorism in the new millennium. An historical analysis of political policing examines the reasons for framing activists as domestic extremists and the murky term ecoterrorist.

In *Discipline and Punish* (1977) and *Power: Essential Works of Foucault* (2002) Foucault outlines three forms of power—sovereign, discipline and governmental.[2] For this chapter, we will focus on discipline in the context of tactics of law. If we understand sovereign power as associated with the law (of the land), then disciplinary power is associated with the "regulation of the actions of individual bodies in schools, barracks, and the institutions of an emerging state" (Death, 2010, p. 17). Technique in this respect refers to the mechanisms, technologies and strategies that shift power through a "body of technical knowledge and practices, [and] a raft of techniques" (Kelly, 2009, p. 44). Thus, techniques become more prevalent within a society, when that society feels threatened to move away from sovereign power into disciplinary power. In other words, those in power impose more restrictions on a perceived threat or danger when it is seen as undermining their position. Before reaching the point of imprisonment, there are levels of disciplinary power where the state begins to control the individual, in order to "organizes an analytical space" (Foucault, 1991, p. 143). In confining an individual's space sets parameters in which their actions monitored. Parameters can be set through either soft or hard power, or a combination of the two. Political Scientists, Joseph Nye (2008) observes that soft power is more than just persuading people to act in a certain way, but also that;

> soft power is not merely influence, though it is one source of influence. Influence can also rest on the hard power of threats or payments. And soft power is more than just persuasion or the ability to move people by argument, though that is an important part of it. It is also the ability to entice and attract. In behaviour terms, soft power is attractive power. (Nye, 2008, p. 3)

Thus, soft power is the art of coercion and persuasion in "getting others to want the outcomes that you want—co-opts people rather than coerces them" (Nye, 2008, p. 95). In Foucauldian terms, soft power would be the technique of the self-; and hard the technique of dominance. This chapter is interested in the use of soft power to execute hard power through legislative amendments, state surveillance, agent provocateurs, and the discourse of ecoterrorism.

Governance and the Organizing of a Political Space for Action

The art of government (as a technique) renders possible the ability to "adjust the multiplicity of men and the multiplication of the apparatuses of production (and this means not only production in the strict sense), but also the production of knowledge and skills in the hospitals, the production of destructive force in the army" (Foucault, 1991, p. 219). Nealon notes that:

> Discipline, as a mode of power is nearly ubiquitous: you wake up to the disciplinary family, consume a breakfast purchased from the efficiency-saturated shelves of the grocery store, and ride the state apparatus—the highway, bus, or train—to school or your job, which in turn is also sodden with the imperatives of discipline: appointments, meetings, tasks, breaks, lunch. (Nealon, 2008, pp. 34–35)

Understood in relation to green governmentality and green capitalism any technique of the self (see Chapter Three) is a series of soft power strategies. Governments and business instigate policies or initiative that reward people—such as green points at the supermarket-to encourage more consumption. Hard power or discipline comes into play if certain groups of the population reject such initiatives in favour of more state led egalitarian policy.

Thus, governments apply a technique of dominance as disciplinary power in order to control and organize the analytical space. In this case, the space is the problem of climate change, and those that reject green capitalism as a solution are organizes into an analytical space that is framed by the governments and media as a threatening space.

With interview data and testimony from those involved in the Critical Mass Bicycle Collective court case, and British Airport Authority (BAA) use of injunction to prevent the Heathrow Climate Camp. This chapter looks at the evolution of political policing. The chapter then moves onto examining the framing of ecoActivists as domestic extremists in the UK, and the nomenclature of ecoterrorism from the USA. By mapping these cultural moments in the recent history of ecoActivism, the chapter identifies a network of political and media discourses, which chime and resonate to produce a space with a heightened sense of protest as a security threat in the media.

Tactic of Law: Legislative Changes and the Criminalization of EcoActivism

For more than a decade USA, UK, Canada, Australia and New Zealand's governments have introduced new legislation or amended current laws that increasingly criminalizes protest. Using legislative systems and spying on activists has become a justifiable tactic to curtail protest. Swathed in a discourse of national security (as Luke (1999) predicted) new laws and increasing application of state sponsored surveillance has the political "intent to drive a wedge between more mainstream and broadly supported organizations (the good) and radical grassroots activists (the bad) seeking to ferment disagreement on one level, and movement splintering on the other" (Salter, 2011, p. 227).

Techniques of dominance are applied legislative changes (tactic of law in Foucauldian terms), and actions such as intelligence gathering and agent provocateurs that led to recasting of the space as a threat to society. Tactics of law is the implementation of "tactics rather than laws, and even of using laws themselves as tactics" so as to "arrange things in such a way that, through a certain number of means, such and such ends may be achieved" (Foucault, 1991, p. 95). As we'll see in this chapter, tactics of law within forms of disciplinary power range from new or amendments to existing laws, political policing including use of agent provocateurs to gather evidence or incite violence and the reframing of ecoActivism under the rubric of domestic extremism—and the neologism of ecoterrorism.

This chapter focuses on two case studies that show attempts by the state to use new laws to curtail and prevent protest—the London Critical Mass cycling collective (2006) and the Climate for Climate Action (aka Climate Camp) near London Heathrow Airport, West London (2007).

Case Study One: Antipodean, Anadarko and Anti-Protest Laws

The New Zealand government amendment to their Crown Minerals (Permitting and Crown Land) Bill (2013) designated any zone around oilrigs and related ships as 'non-interferance zone' to protesters. The Bill's purpose was the protection of oil companies and their rigs. Known colloquially as the *Anadarko law*, after the Texas based oil company Anadarko. The ruling drew criticism from former New Zealand Prime Minister, Sir Geoffrey Palmer. Palmer told the New Zealand media the Bill was an "unlawful limit on the rights of free navigation and free speech" (Inglis, 2013). Stalwart Greenpeace activists feel empathy with Palmer's ideas, as Rainbow Warrior Captain Pete Willcox says

> the ruling goes against New Zealand's right to protest as a fundamental right under democracy. I don't think politicians, in my country [USA] would allow politicians to buy (…) and I don't think in the long run it's going to do New Zealand any good. New Zealand has prideful [sic] history over nuclear issues and other issues. I think it's one of the things that make New Zealand unique. (Interview with Pete Willcox, 2013)

Regardless, the Bill passed as a post-Select Committee supplementary order papers (SOPs) a portal through which to rush law changes. The law gives the Defence force powers to arrest and detain any protesters outside of New Zealand's 12-mile territorial limit. The new law provides a 500-meter exclusion zone around every oilrig in and around New Zealand waters. A protesters or unauthorized personnel entering the zone or getting closer than 500 metres to a ship designated with a non-interference zone will be fined NZ$10,000 (7000 USD). Subsequently, any future activists caught breaking the law or deemed to be interfering with or damaging an oil exploration vessel faces a fine of up to $100,000 for groups or $50,000 for an individual.

Across the ditch in Australia and the states of New South Wales and Tasmania, have both passed laws curtailing and preventing protest to protect logging and mining companies. New South Wales Premier Mike Baird (since retired) passed new anti-protest laws (2016)[3] aimed at the anti-coal-seam gas (Fracking) movements. The new laws gives the police greater powers for more stop and searches, to detain protesters and remove their possessions, and shut down any protest deemed to be blocking traffic, such as blockades or at the road side protests. The law also expands current legislation that anyone interfering with mines to those involved in coal seam gas exploration and extraction sites face a prison sentence of up to seven years. Former New South Wales Chief of Staff to the Treasury, Stephen Galilee and CEO of New South Wales Minerals Council writing for the right leaning *Daily Telegraph* newspaper says "the exaggerated claims of anti-mini activists cannot hide the economic benefits and jobs that projects [Maules Creek mining project] like this will bring" (Galilee, 2013, p. 33).

Tasmania amended their Workplace (Protection from Protesters) Bill—to try to revive the state's logging industry. Seen as an anti-protest Bill, the amendment was for many a tearing up an earlier peace deal between foresters and conservationists. The peace deal enacted was into law just ahead of the 2014 state election. Additionally, an amendment to the Tasmanian Defamation Act 2005 now allows big businesses to sue protesters and have the government remove funding from conservation organizations. The Tasmanian Liberal Party's justification for the changes was that "the Liberals believe that the best way to stop unlawful market attacks by Green groups is not to try and appease them; it's to toughen the law to deter them".[4] The *Sunday Tasmanian* newspaper ask what did 35 years of conflict in the forest achieve. And the answer comes back "Confrontation and conflict associated with this protracted war are ingrained in the Tasmanian psyche and are now part of the local culture". *Daily Telegraph.com.au* ran an opinion piece, claiming logging firm Gunns were being victimized by environmental activists—"Gunns, once one of the best performing companies in Australia, has suffered an even worse fate … after a relentless campaign of eco-terrorism". Australia's former Greens leader, Bob Brown challenged the law in the Tasmanian High Court (October, 2017). The judge ruled the law was unconstitutional after Brown argued they impinged on freedom of political expression. The judge found that whilst protesters were prevented from protesting at the site of contestation it favours corporations. Brown says that "The Tasmanian anti-protest laws are part of the global purchase of democracy by corporations. Corporations lobby weak politicians to legislate against the time-honoured rights of the people"[5]

Across the continent in the state of Western Australia, and the Criminal Code Amendment (Prevention of Lawful Activity) Bill put the onus of proof that damages has not occurred through civil disobedience onto the activists.

The amendment states if an individual or "someone locking themselves as an immovable obstacle, locking themselves onto a vehicle to prevent it from being driven, is interference with the owner's lawful activity can be detained.[6] In effect, the police can arrest anyone suspected of using civil disobedience action such as locking on[7] or tree spiking and it's up to the individual to prove otherwise. Western Australia's Attorney General Michael Mischin told the media that the amendment was "targeted [at] extreme behaviour" adding that in particular fracking had been "hijacked by labels and mantras". Amendments and new laws are not unique to the antipodeans. Like Australia and New Zealand, states across the USA are also passing stringent protest laws while giving more powers to big business and individuals effected by civil disobedience.

Case Study Two: North America and the Onus of Proof on Protesters

USA neoliberalists have been spearheading the reframing of activism as a threat. Such mischaracterization is a common tactic of authoritarian states when those in power label oppositional groups as terrorists, coup plotters, thugs or communists (Stephan and Chenoweth, 2008). In the USA, ten Republican Senators have begun applying a tactic of law strategy to increase penalties against protesters. North Dakota's House Bill 302, was a direct response to the Keystone XL pipeline protests (23 March 2017). The Bill will remove the onus of responsibility on any drivers who may injure protesters with a vehicle. Republican Senator Kempenich said, none protesters could be falsely caught up in the action and, "if something had happened, they'd wind up being accused of it" (Agrawal, 2017). Tennessee have proposed a similar Bill, SB944 (filed by Sen. Bill Ketron (R-Murfreesboro) amending current legislation with a proposal that "a person driving an automobile who is exercizing due care and injures another person who is participating in a protest or demonstration and is blocking traffic in a public right-of-way is immune from civil liability for such injury" (Bado, 2017). In Iowa, lawmakers passed a Bill that meant any protester blocking or restricting traffic faces five years in prison and a fine of up to $7,500 USD. Minnesota is proposing a new law to allow the state to sue protesters for the cost of policing and providing security at demonstrations. Washington state, Republican Senator Doug Ericksen (R-Ferndale) has ambitions to propose a new law that further "criminalises illegal protest aimed at causing economic damage; the unlawful disruption of transportation and commerce and treble damages against funders and organisers" (Luperon, 2017). Virginia State proposes an increase in penalties if protesters fail to disperse from an action under police or authoritative direction. Protesters that refuse to disperse face a fine up to $2,500 and 12 months in prison. North Carolina will make it a criminal offence for a protester to "threaten, intimidate, or retaliate against a present or former North Carolina official in the course of, or on account of, the performance of his or her duties". Colorado, Republican Sen. Jerry Sonnenberg is proposing a law similar to the Anadarko case by reclassifying the law that those interfering with oil and gas equipment charged under a class six crime. Class six offences carries up to an 18 months prison sentence and fine of up to US$100,000. The current position is a misdemeanour.

USA and the Rise of Ecoterrorism Laws

Ecoterrorism is a neologism that began in the 1980, by former Sierra Club member Ron Arnold. Arnold, head of interest group the Wise Use Movement coined the term ecoterrorism. His first book, *Ecology Wars: Environmentalism as if People Matter* (1989) sets out how business not governments should control the American wilderness. Left to governments, national parks and wilderness areas are prone to political lobbying and influence by environmental and conservation organizations; and that "American industry has a moral obligation to protect itself from environmental attacks" (Arnold, 1987, p. 21). Industry needs a role as according to Arnold, if left unchecked environmentalism will become a "new vanguard for a new society in the form of eco-terrorism will become more widespread" [sic] (Arnold, 1987, p. 25).

In June 1998, Arnold's ten-year campaign to label ecoActivists, eco-terrorists culminated in his involvement for a proposed *Eco-Terrorism Prevention Act* (2004). Conflating ecoActivism with animal rights movements, he gave evidence to a Crime Subcommittee of the House Judiciary Committee, Washington. Arnold speaking as the Executive Vice President of the Centre for the Defense of Free Enterprise (1987), argued that

> Eco-terror is a crime committed to save nature, that generally takes the form of equipment vandalism, but which may include package bombs, blockades using physical force to obstruct workers from going where they have a right to go, and invasions of private or government offices to commit the crime of civil disobedience (…) I'm stating that there is no difference between eco-terrorism and animal rights terrorism. And there evidently has been some dispute about that difference. The perpetrators are, in large part, the same people. (Arnold, 1987, p. 54)

Furthermore, Arnold attempted to publically link the ideologies of ecoActivists with Animal Rights collectives such as the Animal Liberation Front, extreme religion groups and acts of terrorism within the rubric of activism. Arnold notes in his 1989 text that; "The adherents of wildernism [sic] are convinced of their moral and ethical superiority, are blind to reason on questions of dogma, and feel they have an exclusive hold on the truth. It all adds up to religious behaviour, and one does not expect objective rationality from religious behaviour (Arnold, 1987, p. 44). Whilst Arnold's attempt at a tactic of law never came to fruition, his ideology steams from the climate skeptic right of American politics.

The USA isn't the only country to apply the term ecoterrorism to ecoActivism, with numerous examples from the UK, and Australia. In Australia, activist Jonathan Moyles is described by the press as "tepee dwelling *eco-terrorist* Jonathan Moylan, who admitted creating a false ANZ press release which briefly carved $340 million [AUD] off the value of Whitehaven Coal". Australian National Party Member of Parliament, George Christensen has referred to Greenpeace Australia as "professional eco-terrorist group Greenpeace".[8] Christensen, in the Australian parliament referred to environmental activists as "extremists pretend to be environmental activists". Debating on the potential damage to the Great Barrier Reef if a mega mine went ahead in north Queensland, Christensen says, "Because they make this false connection to the Great Barrier Reef their gross tactics are accepted by many. But holding the reef to ransom is like putting a gun to society's head and saying give me what I want or the reef gets it. That is an act of ecoterrorism. Their lies, misinformation, slander and the frivolous legal action attacking a company for the sake of furthering an ideological cause can only be described as terrorism if you look at the criminal code"[9].

Case Study Three: United Kingdom and Conflating EcoActivism with Terrorism

UK governments have successfully amended and introduced new laws whilst increasing state surveillance practices against ecoActivists collectives. The Criminal Justice and Public Order Act (1994) originally targeting the new age traveller and rave movements by preventing the gathering of groups, setting up on illegal protest camps, gathering with sound systems and general public order abatement laws that are still used today. The Serious Organized Crime and Police Act (SOCPA (2005) prevents protests being held in central

London without a permit. Previously, no permit was required. Under section 132 [s132] of the Act, it is an offence to organize or take part in a demonstration in a public place within the designated area. A designated area is defined as "no point in the area so specified may be more than one kilometre in a straight line from the point nearest to it in Parliament Square".[10]

S.O.C.P.A was repealed and replaced with the Police Reform and Social Responsibility Act (2011)[11]. PRSRA imposes tighter restrictions on protest making it a criminal offence to protest or set up a protest camp in or around the Houses of Parliament and the Palace of Westminster, without a permit. The Act states that police officer or "authorised officer who has reasonable grounds for believing that a person is doing, or is about to do, a prohibited activity may direct the person to stop" [s.143]. Prohibited activity includes, playing "amplified noise equipment in the controlled area of Parliament Square" including a loud speaker or loud hailer; putting up a tent, or "structure that is designed, or adapted (solely or mainly) for the purpose of facilitating sleeping or staying in a place for any period; any sleeping equipment to sleep over night; and the prohibited activity ban can be up to 90 days" [s.143]. The new law was thought to prevent a reoccurrence of the protest camp set up on Parliament green in 2010 and the Occupy London Stock Exchange (OccupyLSX) protest camps outside St Pauls' Cathedral but originally intended to be outside the London Stock Exchange (2011).

PRSRA also bolstered the UK's Terrorism Act (2000). Part five, "Counter Terrorist Powers" included a change to police powers of arrest, including a "constable may stop and search a person whom he reasonably suspects to be a terrorist to discover whether he has in his possession anything which may constitute evidence that he is a terrorist" [s43]. London Metropolitan Police Commander, Peter Broadhurst describes the rule changing meant "should individuals or small groups seek to take mobilization outside of lawful protest they will be dealt with robustly using terrorism powers" (Vidal and Pidd, 2007). Section 58 and 58a of the Terrorism Act allows police to detain anyone taking photographic or electronically recordings that could be later used for terrorism reasons.

So far this chapter has set out ways that hard power has been applied by governments, interest groups, and lobbyists through a tactic of law to curtail and criminalize some forms of political participation. Using the law and creating new laws that increasingly criminalise protest and activist movements is a growth area within environmental politics—The influence of right-wing republicans in the US, Australia and UK. Political language frame ecoActivism and protest are a threat to both economic stability and national security is a common theme. Techniques of dominance, such as anti-protest laws aim to deter people from joining protest movements. However, not all disciplinary challenges and amendments have been to the benefit of the state or big business. This chapter will now look at two early case studies to show how activists can use such techniques to their advantage. The first example is the case of London Metropolitan Police's trying to shut down Critical Mass London by applying the then SOCPA law. The second example comes from the British Airports Authority attempt to stop the 2007 Camp for Climate Action near Heathrow Airport, West London.

Reaching a Critical Mass

The London Metropolitan Police attempted to make the London Critical Mass cycling collective apply for permission under the SOCPA law. Section 132 of S.O.C.P.A states that any public procession would require advice notification to the Met Police on the basis that it demonstrates, "support for or opposition to the views or mobilizations of any person or body of persons … to publicise a cause or campaign or mark or commemorate an event" [s132]. Additionally, organizers of any public procession must deliver notice, "to a police station … in the police area in which it is proposed the procession will start" [s.132]. Prior to SOCPA, there were no legal requirements for the monthly rides to have a permit. Under s132, without a permit, the rides would be illegal, and participants subject to arrest. Critical Mass stalwart, Des Kay explains it wasn't as simple as the police said, as "The police wanted to know where the route was, who the organizers were, neither of which we can give them, as there's no organizers and we don't have a route. I mean we haven't a clue where we go when we start out" (interview with Des Kay, September 2010).

London's CM or the mass as it's more colloquially known is a celebration of cycling. Cyclists and skateborders' gather on the last Friday of the month under Waterloo Bridge, at London's South Bank. The Mass is a non-hierarchical collective, whose origins are in the "long standing roads protest movement … to create a powerful force as they demand that the government build less roads, [and] fund better public transport" (Anon in Carlsson, 2002, p. 69). The first mass in London took place two years after the inaugural mass in San Francisco, USA (1992). Today around the world, there are over 300 Critical Masses meeting on a monthly basis, with thousands of participants (Anon in Carlsson, 2002, p. 69). London's Critical Mass continues to gather on the last Friday of the month, under Waterloo Bridge, in central London.[12]

Critical Mass bicycle rides are "no protest movement (…) instead, riders have gathered to celebrate their choice to bicycle, and in doing so have opened up a new kind of political space" (Carlsson, 2002, p. 5). Like other collectives, there is no one definition and each participant can "offer a perspective, a manifesto, and a purpose" (Carlsson, 2002, p. 7). The Mass creates its own system of representation and affiliation that links with the wider environmental activist movement—such as a "die-in", "corking", or "swarming".[13]

Figure 4.1. Police stop critical mass bicycle ride in London. Image courtesy of the author.

Des Kay, a former member of CND and ardent cyclist volunteered to front up in court against the Met police's claims. Friends of the Earth (FoE) provided legal support through their newly formed Rights and Justice Centre. Friends of the Earth made it clear to Kay they were willing to support him, but we're unable to represent Critical Mass for fear of biasing the case. Des explains how FoE first contacted him to ask if he'd be willing to "contest this in court" and Kay explains they "we don't really want any of our members, because it would look like it was prejudiced and they said would you like to do it and I was quite happy" (interview with Des Kay, September, 2010).

Phil Michaels, Head of Legal at Friends of the Earth Rights & Justice Centre was Kay's legal representative in court (FOE Press Release, 24 May 2006[14]). Kay explains that Michael's defence was prepositioned on the basis that the monthly rides were neither funeral or procession that he "argued in terms of it being a procession" (interview with Des Kay, September 2010). Michael's argued on behalf of the rides, that the new law meant that any procession that was commonly or customarily held then they could go ahead, as Kay says:

> Michaels argued over various things like if there was a funeral procession or if there was various types of processions would they need a permit. Whereas with the mass, it fell between the exceptions of the law that it was commonly or customarily held, and we ended up winning the case on those grounds. (Interview with Des Kay, September 2010)

As the monthly rides were not subject to the advance notification requirements [1] because they fell within an exception for processions that are "commonly or customarily held", then "it's ok to have a procession if it's customarily held", and as the London Mass "was customarily held because it had been running for thirteen years", and "not really a procession anyway (...) so that's how we ended up legitimised" (interview with Des Kay, September 2010). The legal ruling by the Divisional Court (of two judges) originally found that:

> an unbroken succession of over 140 of these collective cycle rides, setting out from a fixed location on a fixed day of the month and time of day and travelling, albeit by varying routes, through the Metropolitan Police area, cannot by now sensibly be called anything but common or customary.

The Metropolitan police appealed and won with the Court of Appeal rejected the original judgment by a majority of 2 to 1. Michael's and Kay took the case to the House of Lords, as Kay explains,

> When they [Met Police] appealed, their appeal was that the Mass although it's commonly and customarily held, because it's not the same route each time, then it can

be a procession, then that argument wouldn't be valid you see. (Interview with Des Kay, September, 2010)

Michaels' in a Friends of the Earth Press release explains further,

> We are in an unusual situation where the police's appeal has been successful even though, over two separate hearings, three judges have found in our favour and only two have found against us. Although the result is disappointing we are encouraged by the strongly worded dissenting judgments in our favour and hope that we are able to take this important case to the House of Lords.[15]

Friends of the Earth and Kay undeterred after the Court of Appeal found that the monthly rides were not exempt from the requirement to give advance notice under the Public Order Act.

> So we then took it to the House of Lords [21/5/2007], and using the same argument, there was five Law Lords and they unanimously supported our argument (…) [we] mentioned the fact that it's not really a procession anyway you know so they were quite happy with that whole arrangement really and so that's how we ended up legitimized. (Kay interview, 2010)

Critical Mass London will celebrate its 25 anniversary in 2019, and riders continue to meet on the last Friday of the month, under Waterloo Bridge on London's South Bank. Mass's continue to spring up around the world today and are an integral part of environmental activists movements.

Criminalizing Protest to Protect Corporations: The Injunction and the Activists

The Terrorism Act of (2000) and the Protection from Harassment Act (1997) were used in combination between the police and British Airports Authority in an attempt to restrict a protest camp near London Heathrow Airport. British Airports Authority (BAA) owns London Heathrow airport, and when the Camp for Climate Change chose Heathrow airport as the target of their third Climate Camp, BAA used anti-stalking laws to stop any direct action in or around the airport from taking place. The British Airports Authority (BAA) in an attempt to stop activists from the third Camp for Climate Change (2007) carrying out direct action at Heathrow Airport, applied for an injunction using the Protection from Harassment Act (1997). The Act primarily focuses on anti-stalking legalization that includes restraining and abatement orders.[16] BAA's choice to use the law was to apply for an "injunction that typically ban activists from demonstrating against companies or going near their premises" (Lewis and Evans, 2009, p. 10). BAA took out an injunction against numerous individuals and movements. The injunction aimed to prevent action-taking place within a certain radius of the west London airport.

BAA's injunction was supported by the Police's use of the Terrorism Act [s44] stop and search powers. Section 58 was applied to two activists cycling near the airport. They were held for 30 hours under [s58] for the offence of "collects or makes a record of information of a kind likely to be useful to a person committing or preparing an act of terrorism, or (b)he possesses a document or record containing information of that kind".[17] A record can be a photograph, note, or any form of electronic record. The injunction and use of s58 of the Terrorism Act, in effect banned fifteen environmental groups from setting up a camp at "or in the vicinity of Heathrow and from carrying items including spades, saw as, ropes, cables, aerosol cans, balloons, whistles and loudhailers" (Hickman, 2007). In addition, four individuals were named in the injunction from Climate Camp, Plane Stupid, and HACAN. The injunction stipulated that "protesters would be able to gather at three protest points on the outskirts of the airport providing that they do not exceed an as yet unspecified number, give their names, car registration plates [sic] and advance notice. They would not be allowed to use any megaphones, klaxon's or sirens or go within 100 meters of any airport operation" (Hickman, 2007).

However, the injunction was so broad that it would have "prevented the three to four million members of the National Trust, including the judge hearing the case, from approaching Heathrow" (Price, Saunders, and Olcese, 2014, p. 169). Groups effected included, Airport Watch, Greenpeace, Friends of the Earth, the Royal Society for Prevention of Birds, the National Trust, the Campaign for Protection of Rural England, the Women's Institute, Rising Tide, Plane Stupid, HACAN, Camps for Climate Action, London Greenpeace, Transition Heathrow and many more. With such a broad range of interest groups implicated, the injunction had the opposite effect for BAA, and encouraged people to attend the weeklong climate camp. Steve, an activist helping to put the camp together explains:

> It was a massive piece of luck of the injunction. The injunction worked in favour of the camp, as you couldn't wish to find a better own goal than that. The National Trust, hang on a minute, they're militants, the Queens National Trust, and the Queen, hang on a minute. And that started people. It was pure luck really. (Interview with activist Steve, 2009)

With a ground swell of support the third Camp for Climate Action took place legally at a nearby sports ground owned by Imperial College London. In hindsight, many activists spoken to, felt that the Heathrow camp was a tipping point for the movement. Activists and likeminded individuals saw first-hand how governments and business were using laws as a tactic to discipline and curtail not just ecoActivism but all forms of protest. When governments curtail individual rights for the benefit of multinational companies, it has a reverse effect on the public. Airport authorities, the police, the government wanted to try and discredit it beforehand. In presenting the camp as problematic needing an injunction to the press and therefore the public, a narrative of potential violence and security concerns, was all a tactic of law to. Instead of preventing protest at Heathrow, it encouraged more people and more action that is public. HACAN chair Jon Stewart (JS) explains:

> they [police and BAA] wanted make sure as few people as possible came to it [the camp], beforehand. I think what they wanted to try and do with this sort of headline is say to younger and older people who might both have been that involved, oh I like the sound of that, oh, oh my god, I won't go. (Interview with JS, September 2011)

Media coverage of the injunction and awareness of how many people would be affected also worked to the climate camps' benefit. Coverage of the injunction and those affected made for several news headlines. *The Guardian* newspaper were blunt in their reporting supported the protest with headlines focus in use of laws as a tactic to curtail protest with "Police to use Terror laws on Heathrow Climate Protests", and "Very Modern Protest: a Rainbow coalition of angry residents, greens and a local MP will defy BAA to join next week's Heathrow camp" (Vidal and Pidd (2007). *Daily Mail* newspaper reported *BAA wins injunction to ban climate change protest.*[18] In reality, the national press coverage had made more people aware of the camp that might not have been prior to BAA injunction.

The vast swathes of organizations, groups, movements, collectives and individual that may have been affected raised interest in the camp at a global level. Climate Camp facilitator, Steve says, "The week before there were hundreds of people, we were getting over a hundred people turn up at weekly meetings—which was unbelievable" (interview with activist Steve, August, 2009). Activists purposefully chose Heathrow as a protest campsite, to tap into a groundswell of interest over climate change. It was as Steve says "a moment in time really, because it was when everybody wanted to get involved with the climate camp and in fact Heathrow and Kingsnorth has become a bit like the 1966 World Cup Final, where there were only about 60,000 people there and about 300,000 people say they were there" (Interview with activists Steve, 2011).

Plane Stupid activists JS (2013) sees the injunction as a panic measure against large numbers of people supporting the camp, he says;

> What worried them [BAA] at the Climate Camp more than others, because potentially this was really going to challenge the power of the state. The campaign might have won, but also if people were going to be running onto runways that are challenging some and that's why, I suspect, there was more concern prior to the Heathrow camp than others. That's why also, I think personally, going to Heathrow was a brilliant idea because it did exactly that, it was genius to go there at that time. (Interview with John Stewart, 2011)

Following the media coverage and comments from organizations such as the Women's Institute and National Trust, BAA revised their application and withdrew the injunction against individuals. The High Court heard that BAA wished to limit any designated protest sites to three areas. Timothy Lawson-Cruttenden, BAA's lawyer, asked the court for an order designating three public protest areas on the Heathrow site where the police will "regulate behaviour to ensure lawful activity takes place" whilst preventing protesters carrying out "unlawful" activity—obstructing the operation of the airport or its passengers—elsewhere at the airport.

BAA use of legislation (as a tactic of law) and, policing of the climate camp protest centred on the protection of the corporations and security of their ongoing business. Such support of neoliberal interests by the police to protect business is one of the ways tactics of law has led to the "the criminalisation of peaceful activism under the rubric of 'anti-terrorism' is an escalating trend linked directly to corporate co-optation of the national security apparatus".[19] In applying, a tactic of law it's often combined with state sponsored surveillance techniques of dominance that gather intelligence via metadata, undercover policing, and agent provocateurs.

Not all tactics of law and technique of dominance are successful at criminalizing ecoActivism and in the case of BAA and Climate Camps (2007 and 2008); it was a strategy that backfired.

All of these examples and many more show how hard power is an open and transparent way of using current laws to curtail protest. However, soft power as a more nuanced technique of dominance can also include surveillance. Applying a capture surveillance approach—intelligence gathering through data collection, spying, eavesdropping on phone calls and conversations, data-mining emails, social media accounts etc. has increasingly become a key strategy of the political policing. The following section looks at how the UK government financed several political policing units to gather intelligence on ecoActivists in particular, as justification for the introduction of new laws.

Secret Police: Political Policing as a Technique of Dominance

Political policing and the extent the UK were using undercover intelligence gathering strategies were revealed in 2015. Lord Justice Pitchford chaired a Parliamentary inquiry into undercover policing practices. The Pitchford Inquiry has three objectives. Firstly, to establish the facts, motivation, and scope of undercover policing, and the role undercover policing plays in prevention and detection of crimes. Secondly, to investigate processes, governance, justification, authorization, training and selection and awareness of undercover policing within the government. Thirdly, to draw on witness testimony to assess the future of undercover policing in the UK (some of which is from the same activists interviewed for this book)

The inquiries initial findings reveal that political policing had taken place against numerous groups' individual political activists; trade unions and members of the Blacklist Support Group; Good Easter Hunting saboteurs; the McLibel support group, London Greenpeace; Climate Camp Legal Team; artivists from the Clandestine Insurgent Rebel Clown Army; Genetic Engineering Network; AWE Aldermaston, Reclaim the Streets.[20] Other findings revealed that some police officers had taken the identity of deceased children and adults who had died at an early age. In court testimony heard that some police officers had used sex to elicit information from some activists. The enquiry confirmed what many from new social movements, collectives, individuals, researchers, journalists and others already knew. Taking a stance against a politically led policy would subject individuals and groups to state sponsored intelligence gathering practices. At the time of writing, the inquiry is still to give its final report.

Political Policing in the UK

The Special Demonstration Squad (SDS)—a political policing unit—began in response to rising civil disobedience in the late 1960s. SDS was part of the London Metropolitan Police Special Branch, also known as the *Hairies*.

Hairies, a nickname given to police officers in the unit because they had to grow their hair long to fit in with the hippy movement. SDS was the first undercover police unit—formed in recognition for better intelligence, equipment and training of public order work. In light of the anti-Vietnam protests, a growth in Campaign for Nuclear Disarmament (CND) memberships, non-violent direct mobilizations and civil disobedience emerging civil rights, feminism and environmental protests movements of the late 1960 and early 1970.

Mid 1990 and the SDS shifts its focus from intelligence gathering on civil rights and anti-nuclear activists to the environmental activists movements. EcoActivism's heightened mobilization by the collectives Reclaim the Streets, Clandestine Insurgent Rebel Clown Army and the anti-roads protests, led SDS to announce it was "changing its priorities to concentrate on environmental activism" (Monbiot,[21] 1996). EcoActivism became the new target of undercover policing on the assumption that "environmental activists might be preparing for suicide attacks on road builders" (Monbiot, 1996), despite there being "no apparent rise in the threat from radical politics to explain the change in gear. The nature of protest was, as ever, changing ... giving way to more imaginative non-violent direct mobilization protests, increasingly facilitated by the use of the internet" (Lewis and Evans, 2013, p. 199).

Political policing as a technique of dominance increases in the new millennium. Disbanded to make way for three new intelligence-gathering units, the SDS was reinvented as the National Public Order Intelligence Unit first (1999) (NPOIU), the National Extremism Tactical Coordination Unit (2004) (NETCU), and National Domestic Extremism Team (NDET) (2005) a national policing unit within a Specialist Operations business group of the Metropolitan Police Service (MPS). In addition, the Counter Terrorism Command (CTC, 2006) and Protection Command and Security Command where formed to gather intelligence to protect big business. Under the guidance of the Association of Chief Police Officers (ACPO)[22] political policing was then undertaken prior to and during large-scale gatherings for counter terrorism and security planning purposes. The numerous units gathered intelligence and provided advice to aid businesses on security. New police powers enabled greater monitoring, intelligence gathering, and restriction on planned protest in order for appropriate policing to take place (Hattenstone, 2011). With so many different units and political policing strategies there reaches a point that "by 2005, both the SDS and NPOIU had full fleets of spies planted in protest organizations. With two rival units, it was sometimes impossible to work out who exactly a police spy was working for" (Lewis and Evans, 2013, p. 211).

ACPO oversaw the merging of the three units (NPOIU, NDET and NETCU) in 2011, and "ratified the decision for the Metropolitan Police Service (MPS) to become the lead force for the National Domestic Extremism and Disorder Intelligence Unit (NDEDIU)".[23] NDEDIU were tasked to "provide intelligence on domestic extremism and strategic public order issues in the UK. Police will always engage to facilitate peaceful protest, prevent disorder and minimize disruption to local communities. Where individuals cross over into criminality and violence, the police will act swiftly and decisively to uphold the law".

Former M15 officer, Annie Machon was openly critical of the new groups. By targeting protest is for Machon tantamount to "effectively spying on individuals for having opinions that the government disliked, using a definition of subversion that was so loose that it included anyone who was exercising their democratic right to protest" (Lewis and Evans, 2013, p. 204).

With the National Domestic Extremism and Disorder Intelligence Unit (2004) (NDEDIU) leading the way, the nomenclature of *domestic extremism* becomes a new label to define environmental activism and protest in general.

Domestic Extremism as a Technique of Dominance

Domestic extremists can be anyone, but the police chose to focus specifically on

> Two groups: hard-core animal rights activists behind sometimes nasty campaigns against vivisectionists and a group defined as "environmental extremists", taken to be a loose reference to the anti-roads movement, campaigners against war, nuclear weapons, racism, genetically modified crops, globalisation, tax evasion, airport expansion and asylum law, as well as those calling for reform of prisons and peace in the Middle East. (Lewis and Evans, 2013, p. 203)

The head of the National Public Order Intelligence Unit (NPOIU), Barry Moss, coined the UK definition of a domestic extremist. A general definition of a "domestic extremist" is anyone engaging in, "political activity involving criminal acts of non-violent direct mobilization in furtherance of a single issue campaign" (Lewis and Evans, 2013, p. 202). Environmental activists could be understood as domestic extremists, "with no legal basis or agreed definition, the phrase could be applied to anyone police wanted to keep an eye on. In the most general terms, it was taken to refer to political activity involving "criminal acts of non-violent direct mobilization in furtherance of a single issue campaign" (Lewis and Evans, 2013, p. 202). Domestic extremists can also be someone or something that police officers identify as a group, community or individual who wanted to "prevent something from happening or to change legislation or domestic policy", often doing so "outside of the normal democratic process" (Lewis and Evans, 2013, p. 202).

Green Party Peer, Baroness Jenny Jones of Moulsecoomb, was characterized as a domestic extremist for her support of the environmental group, Critical Mass and the Occupy Movement and found she had been added to the National Domestic Extremism and Disorder Intelligence Unit (NDEDIU) list of domestic extremist in 2011. Jones, writing in *The Guardian* newspaper explains why she found herself on the database: "The supposed point of this database, which is managed by the Met, is to gather intelligence from police forces, counter terrorism units, industry and open sources about domestic extremism threats, of which I am apparently one" (Jones, 2014). Others on the list included, Ian Driver, a Green Party Member for Thanet, Kent; and a Brighton pensioner who won a legal challenge against the list. Following a request by the London Mayor's Office for Policing and Crime (MOPAC) to define domestic extremism given there's no legal definition, a revised definition of domestic extremists was agreed as "domestic extremism relates to the activity of groups or individuals who commit or plan serious criminal activity motivated by a political or ideological viewpoint".[24]

Applying a discourse of extremism as a technique of dominance is an attempt to apply a one size fits all model to justify amending public order policy. Scooping up all forms of activism under the nomenclature of "domestic extremists" gives justification to increase usage of tactics of laws. In reframing, or more precisely to recasting (borrowing Luke's term (1999)) of ecoActivists as extremists provides the wherewithal to justify increasing usage of tactics of law through a public order defence.

Supported with intelligence from political policing on ecoActivists leads to a re-framing of as something that "endangers national security" (Luke, 1999, p. 125). Such progressive policing supported by state surveillance has grown over the last 25 years. Applying the label eco-terrorist or domestic extremists to frame activists' movements leads to public opinion that interpolates activism as extremism. Luke's point that "environmental resistance can even be recast as an act of national security" is beginning to manifest as a truth.

Political Policing: PC Mark 'Stone' Kennedy

PC Mark Kennedy was the most prominent of all the undercover police officers, when he was revealed by activists, and dishonored by the police service and accused of going native. Kennedy embedded with activists that included personal and sexual relations in order to pass on intelligence on UK and Europe activists movement was revealed after the Radcliff on Soar protest. PC Mark Kennedy was an officer in the National Public Order Intelligence Unit (NPOIU) for eleven years. During his time at NPOIU, he infiltrated numerous environmental activists' movements to gather intelligence so appropriate policing could take place (Hattenstone, 2011). Quickly gaining trust within the movement he was bestowed the nickname Flash because of access to money (mainly from overtime pay from the police force), a van, and expensive climbing equipment. Nottingham based activists exposed Stone when a girlfriend found a passport with his real name—Mark Kennedy (2010). Having infiltrated the movement for over ten years, Kennedy was involved in and had "been around a lot of movements—Earth First, anti-fascism, Dissent (you may remember him as one of the organisers of the driving teams at Stirling)" (Indymedia, 2010). Kennedy's exposure led to many more undercover police being revealed within the movement.

PC Mark Kennedy was not the first police officer embedded with environmental movements, but he was the first to confirm what activists had suspected for years that—state sponsored police surveillance, aka political policing was common practice. Exposing Kennedy soon led to further revelations of more officers had been working undercover in activist movements since the 1990s.

At least four other undercover police officers were unveiled as fake activists. PC Jim Boyling a.k.a Jim Sutton was "mainly tasked with infiltrating a raucous collective known as Reclaim the Streets, notorious for occupying roads and hosting impromptu street parties. It was an anti-capitalist, green movement against the dominance of cars" (Lewis and Evans, 2013, p. 172). PC Lyn Watson was revealed to have infiltrated the anti-nuclear movement in 2003, the same year that Kennedy was attached to the environmental activists' movement. Watson "first surfaced in 2003 at a protest at Aldermaston, Britain's nuclear weapons centre", and her "abrupt disappearance in 2007 caused concern about her welfare rather than suspicion of her role" (Syal and Wainwright, 2011). A fourth police officer PC Mark Jacobs worked inside the Global Justice Protest Movement and anarchist groups (2005–2009), first as an activists at the G8 summit (2005) and later the London based anti-capitalist protests (2009).

Kennedy's was by far the biggest media story, when it was revealed his intelligence gathering about plan to hold a weeklong direct action campaign and protest camp at the Radcliffe-on Soar coal fired power stations,

Nottinghamshire, England (April, 2010), was to be used to convict 20 activists with conspiracy to commit aggravated trespass. The power station, owned by energy company E.ON UK was due to be the site for a weeklong non-violent direct action protest camp (2010). Two hundred activists helped stage a protest with Kennedy a key figure in the organization and planning of the camp. Unbeknown to those taking part, each morning Kennedy would gather intelligence from the daily assemblies, by recording conversations on a modified wristwatch. Kennedy then left details of protest plans on the seat of his vehicle ready for his police handler to take and photocopy. Two days before the planned mobilization (April 11) some activists became concerned over what they felt was an increase in police patrols around the perimeter of the power station. Kennedy volunteered to check it out, and returning from his reconnaissance trip lied to the group that there was no police presence, when in fact there was.

April 13th, 2010 and uniformed police raided the assembly point for the action, Iona School in Sneinton, ten miles from the power station. Police arrested 114 people and 22 charged with conspiracy to commit aggravated trespass. All 22 were later acquitted by the High Court in London. The exposé of political policing and the Radcliff on Soar court case led to a public enquiry held at the Houses of Parliament. Kennedy's actions were symptomatic of wider political policing strategies, and government policies towards activist, protests, radical movements, some new social movements and certain individuals, which as we will see is a means to recast ecoActivism as an act of deviancy, to justify the curtailing and limiting of protest within a democratic society and the reframing of activism as domestic extremism, all of which can be understood as power relations between activism and the state.

Tactic of Law from police intelligence gathered from surveillance has led to numerous arrests and acquittals in the past ten years. Twenty-nine ecoActivist were sentenced and fined for hijacking a coal train near the Drax power station, Yorkshire (2008). One hundred and fourteen ecoActivists charged and later acquitted with "pre-emptive trespassing" at the Radcliffe-on-Soar cold fired power station (2009). Of the 114, twenty were (Moses, 2013)[25] "convicted of conspiracy to commit criminal damage and aggravated trespass" all charges later dropped and protesters cleared (2011). Emma Sheppard, one of the activists was given a two-year prison sentence for endangering police officers lives with a homemade traffic "stinger" device. Other arrests and convictions include nine Scottish activists (Known as climate nine) were charged with breach of the peace during a mobilization at Aberdeen airport (2009).

Each of the nine were given fines ranging from £300 too 700 each; Twenty one protesters were charged with aggravated trespass for an mobilization at West Burton power station, Nottingham(2013). The 21 were given community service orders. The case gained media attention, when the parent of one protester raised 64,000 signatures with an online petition against West Burton owners, EDF energy's threat to sue the activists for five million dollars. Petition Signatories included evolutionary biologist, Richard Dawkins, actor Mark Ruffalo, author Naomi Klein, and cultural theorists, Noam Chomsky. *The Guardian* reports, "Several hundred apparent EDF customers claimed they had switched their energy provider. EDF said it dropped the civil mobilization as part of a settlement with the protesters, who agreed to accept a permanent injunction preventing them from entering EDF sites" (Carrington, 2013).

Agent Provocateurs and Political Policing at G20

One final political tactic to look at is the is the training and engaging of agent provocateurs. Certainly not a new tactic of law by the state to gather information and arrest, and like the undercover policing, many activists understand that undercover police and intelligence gathering officers are agent provocateurs. Agent provocateurs are police officers disguised as activists, whatever those disguise takes and insight trouble or encourage others to break the law to cause arrests. Agent provocateurs have been present during the Critical Mass court case, the G20 Climate Camp and anti-capitalism mobilizations. Lewis and Evans (2013) note with that with a:

> a small numbers of activists at some meetings, agent provocateurs were seen on at least two occasions, there were as many corporate spies at meetings of the small group as genuine activists. The number of real activists in the meetings had by then dwindled to between five and ten, and sometimes fewer. The spies had an incentive to put more effort into building their credibility in the group and so often attended more meetings than bona fide campaigners. Remarkably, agents working for McDonald's were effectively helping to hold its archenemy London Greenpeace together. (2013, p. 71)

The G20 protest in the UK (2009) is synonymous with the death of newspaper seller, Ian Tomlinson after an altercation with Police Constable (PC) Simon Harwood. At a mobilization during the Group of 20 (G20) protests, in London (2009) activists, journalists and a Liberal Democrat Member of Parliament (MP) were witness to agent provocateur activity.

Critical Mass activists DK (2013) says, "There's clearly agent provocateurs you know involved and that was never more clearly demonstrated than at the Bishopsgate climate camp". The climate camp was part of a larger global justice/anti-capitalist protest at Bishopsgate (2009) and surrounding streets in the City of London. Mobilization took place across the City of London, and a temporary climate camp set up outside the alternative energy exchange. Four different marches converged outside the Bank of England. Police applied the tactic of kettling (holding protesters in a confirmed space and slowly like a kettle tensions reach boiling point). Many protesters and individuals testified to seeing agent provocateurs attempting to incite violence during the containment. One individual was throwing bottles at the police in an attempt to provoke a reaction. And when "people confronted him [police officer], he walked out and he was able to walk straight through the police cordon when the rest of the people were kettled … because it was obviously, just some agent provocateur you know, why else would they let him go really" (interview with MS activists, 2013).

Liberal Democrat MP, Tom Brake gave evidence of two people throwing bottles at the police line, only to be able to pass through the line without being questioned. Brake told the *Guardian* newspaper—"When I was in the middle of the crowd, two people came over to me and said, "There are people over there who we believe are policemen and who have been encouraging the crowd to throw things at the police".[26] Giving evidence at a Human Rights Joint Committee hearing on the G20 protest, Brake says

> An individual who was encouraging the crowd to push against police lines, throw things against the police lines and who, when challenged by the protesters who said, "We think you're a police officer", subsequently left the police lines and walked through the police cordon.[27]

Conclusion

Governments use intelligence and evidence gathered from political policing and other means to create disciplinary spaces. By generating a discourse of deviancy, that labels activists as domestic extremists and/or ecoterrorism is achieved through various anti-protest laws across the globe. The laws then generate a sense that protest and activism is a threat that needs to be addressed and therefore provides a justification for further laws to limit protest. Such justification is in Foucauldain sense is a move from technique of self to technique of dominance, or persuasion to discipline, to provide governments with the tools to organize activism within a set analytical space boarder by legislation and maintained through political policing. Civil disobedience becomes harder to maintain as part of the democratic process, and provides a public justification to introduce new laws and increase political policing on anyone participating or planning civil disobedience. Tactic of law come to define cultural moments and provide political justification for more legislation. Moreover, in a highly security conscious society governments attempts to "rationalize" a problem like the climate debate, has led to a division between part of the population that agree economics are key to climate change solutions, and those seeking a more socially just solution to become abnormal, and in some cases a threat to national security.

Media coverage support the government through an inferential structure that places ecoActivism as domestic extremists, ecoterrorist at the most amplified point, to become the starting point for using tactics of law to prevent protest centre tactics law (BAA injunction and the Critical Mass court case that act to splinter and disintegrate new social movements. Tactics of law of political policing, agent provocateurs and surveillance all add to the criminalization of protest. From 1997 and the election of a Blair/ Brown New Labour government there was a rapid increase in state surveillance of protesters, activists and general public interested in environmental issues and the right to protest. At the same time, undercover policing, capture and open surveillance, agent provocateurs, and the neologism of eco-terrorism and domestic extremism all feed into an amplification of media representation of ecoActivism as violent, negative, security threat and potential criminal act. First seen in protest reporting in 1970s and 1980s (see Chapter Two), throughout the latter half of the 1990s and into the new millennia, the hushing of the protesters voice by the media and legislation empties out the activists' voice from the climate change debates to be replaced by representations of the state.

Notes

1. See https://www.indymedia.org.uk/en/2010/10/466477.html for full statement.
2. The third category, governmental power, is associated with "techniques employed to achieve biopolitical government of the population ... working through the conduct of conduct rather than ruling directly over territory and bodies" (Death, 2010, p. 17).
3. https://www.parliament.nsw.gov.au/researchpapers/Documents/protests-and-the-law-in-nsw/Protests%20and%20the%20law%20in%20NSW.pdf
4. See https://www.tas.liberal.org.au/sites/default/files/policy/Rebuilding%20the%20Forest%20Industry%20-%20cracking%20down%20on%20protestors.pdf for further information

5. See http://www.bobbrown.org.au/tasmania_s_anti_protest_laws
6. http://www.news.com.au/national/breaking-news/was-antiprotest-law-opponents-to-rally/news-story/6b307b89adb395f037a9c95c737b1204
7. Locking in on is an old school tactic of using bicycle or similar locks to attach the activist to a piece of machinery, tree or whatever is the focus on the protest, and sometimes activists lock on to each other.
8. https://www.georgechristensen.com.au/tag/eco-terrorist/
9. See http://www.openaustralia.org.au/debates/?id=2016-02-08.8.2 for full transcript.
10. http://www.legislation.gov.uk/ukpga/2005/15/section/138 for further details
11. http://www.legislation.gov.uk/ukpga/2011/13/section/143/enacted
12. Setting of around 6:30 to p.m., the mass takes an unplanned route around central London. Riders at the front, rotate of cyclist at the front of the mass that are never leading. There's a few favourite haunts of the London Mass—around the Victoria Monument outside Buckingham Palace or a few laps around Parliament Square in Westminster, the east end of London, Whitechapel and Bow areas and the central shopping areas of Oxford and Regent Street. Cyclists will occasionally visit the site of a cyclist's death from a road accident, often marked with a ghost bike—a white painted bike chained up.
13. The Mass also has its own rituals from 'corking', 'navigating', 'breaking the mass' or splitting the mass' to negotiate the ride. A die-in is a form of protest when activists pretend to be dead, often symbolizing the death of the environment. The die-in is a common tactic of the anti-aviation collectives, and occasionally Critical Mass. Swarming, is a protest tactic, where activists split into small groups ahead of a protest, and then come together as one, from different directions, to create a mass. Corking is when one or more cyclists place themselves on their bicycles in front of any waiting traffic at junctions and roundabouts. Once the Mass has passed, the cyclists move to allow traffic to continue. A navigator is one or more bicyclist at the front of the ride who is followed by the rest of the ride. It is an important part of the ritual of the ride that the navigator always changes; Breaking or splitting the mass occurs when at a red light a large group of riders thins out to create a safer mass. (accessed at http://www.joelpomerantz.com/genlresources/cmglossary.html#redlight
14. For more information see http://www.foe.co.uk/community/campaigns/rights/rights_justice_centre16424.html (accessed November 2009).
15. https://www.foe.co.uk/resource/press_releases/legal_ruling_on_critical_m_21052007
16. http://www.legislation.gov.uk/ukpga/1997/40/crossheading/england-and-wales
17. http://www.legislation.gov.uk/ukpga/2000/11/section/58
18. http://www.dailymail.co.uk/news/article-473456/BAA-wins-injunction-ban-climate-change-protest.html#ixzz4f8dziPtO
19. http://www.theguardian.com/environment/earth-insight/2014/jan/21/fracking-activism-protest-terrorist-oil-corporate-spies
20. https://www.ucpi.org.uk/wp-content/uploads/2015/12/121216-list-of-core-participants-in-receipt-of-public-funding.pdf
21. http://www.monbiot.com/1996/08/29/defective-inspectors/
22. ACPO was disbanded in 2015 and replaced by the National Police Chiefs' Council (NPoCC) (2015).
23. http://www.npcc.police.uk/About/History.aspx

24. https://netpol.org/wp-content/uploads/2014/04/jj-3367-Definition-of-Domestic-Extremism-BHH-new-definition.pdf
25. http://www.theguardian.com/environment/2013/jun/06/climate-protest-courts-uk-timeline
26. http://www.theguardian.com/politics/2009/may/10/g20-policing-agent-provacateurs
27. http://www.publications.parliament.uk/pa/jt200809/jtselect/jtrights/141/9051203.htm

References

Agrawal, N. (2017). “In North Dakota, It Could Become Legal to Hit a Protester with Your Car”, *LA Times Newspaper*. Available at http://www.latimes.com/nation/la-na-bills-protest-criminal-20170201-story.html

Arnold, R. (1987). *Ecology Wars: Environmentalism as If People Mattered*, Bellevue, WA: Free Enterprise Press.

Bado, K. A. (2017). “Bill Would Make Drivers Immune from Civil Liability in Protests”, *USA TODAY NETWORK - Tennessee*. Available at http://www.tennessean.com/story/news/politics/2017/02/10/bill-would-make-drivers-immune-civil-liability-protests/97743390/

Carlsson, C. (2002). *Critical Mass: Bicycling's Defiant Celebration*, Edinburgh: AK Press.

Carrington (2013). “Climate Activists Receive Sentences for Gas-Fired Power Station Occupation”, *The Guardian*, June 7th. Available at https://www.theguardian.com/environment/2013/jun/06/climate-activists-sentenced-gas-fired-power-protest

Darier, E. (1999). *Discourses of the Environment*, Oxford: Blackwell.

Death, C. (2010). *Governing Sustainable Development: Partnerships, Protests, and Power at the World Summit*, Hoboken, NJ: Taylor & Francis.

Foucault, M. (1977). *Discipline and Punish the Birth of the Prison*, London: Penguin.

Foucault, M. (1991). “Governmentality”, in Burch ell *et al.* (eds.), *The Foucault Effect*, Hemel Hempstead: Harvester Wheatsheaf, pp. 87–104.

Foucault, M. (2002). *Power: Essential Works of Foucault*, London: Penguin.

Galilee, S. (2013). “Activists Misleading the Community with Fiction”, *The Daily Telegraph*; Surry Hills, N. S. W., 15 Feb 2013, p. 33.

Hattenstone (2011). “Mark Kennedy: Confessions of an Undercover Cop”, *The Guardian*. Available at https://www.theguardian.com/environment/2011/mar/26/mark-kennedy-undercover-cop-environmental-activist

Hickman, M. (2007, July 27). “Heathrow Puts up Legal Barricades to Keep Away Protesters; Heathrow Security Alert”, *The Independent*. Available at https://search-proquest-com.elibrary.jcu.edu.au/docview/311325622?accountid=16285

Inglis, J. (2013). “New Zealand's Government Sides with Texas Oil Giant, Anadarko, Bans Protests Around Its Oil Right”, *PRI.org*, November 26. Available at https://www.pri.org/stories/2013-11-26/new-zealands-government-sides-texas-oil-giant-anadarko-bans-protests-around-its

Jones, J. (2014). "The Met Turned Me into a Domestic Extremist – with Tweets and Trivia", *The Guardian*. Available at http://www.theguardian.com/commentisfree/2014/jun/16/domestic-extremist-metropolitan-police-spying-elected-politician

Kelly, G. E. M. (2009). *The Political Philosophy of Michal Foucault*, London: Routledge.Lewis, P and Evans, R. (2009). High Court Injunction- the Weapon of choice to slap down protest. The Guardian, 28th October Lewis, P., and Evans, P. (2013). *Undercover: The True Story of Britain's Secret Police*, London: Faber and Faber.

Luke, T. (1999). "Environmentality as Green Governmentality", in E. Darier (ed.), *Discourses of the Environment*, Oxford: Blackwell, pp. 121–151.

Luperon, A. (2017). "Lawmaker Seeks to Criminalize Protests as 'Economic Terrorism'". Available at http://lawnewz.com/high-profile/lawmaker-seeks-to-criminalize-protests-as-economic-terrorism

Moses, K. (2013). "Climate Protests in the UK Courts – Timeline", *The Guardian*, June 7th. Available at https://www.theguardian.com/environment/2013/jun/06/climate-protest-courts-uk-timeline

Nealon, J. T. (2008). *Foucault Beyond Foucault: Power and Its Intensifications Since 1984*, Stanford, CA: Stanford University Press.

Nye, J. S. (2008). "Public Diplomacy and Soft Power," *The Annals of the American Academy of Political and Social Science*, 616(1), pp. 94–109. doi:10.1177/0002716207311699

Price, Saunders, and Olcese (2014). "Movements", in C. Death (ed.), *Critical environmental politics*, London, New York, Routledge: Taylor & Francis Group.

Salter, C. (2011). "Activism as Terrorism: The Green Scare, Radical Environmentalism, and Governmentality, Anarchist Developments in Cultural Studies Ten Years after 9/11: An Anarchist Evaluation", *AVDCS Journal*. Available at http://www.anarchist-developments.org/index.php/adcs/article/view/45 (accessed February, 2012).

Stephan, M. J., & Chenoweth, E. (2008). "Why Civil Resistance Works: The Strategic Logic of Nonviolent Conflict", *International Security*, 33(1), pp. 7–44. doi:10.1162/isec.2008.33.1.7

Syal, R., and Wainwright, M. (2011). "Undercover Police: Officer A named as Lynn Watson", *The Guardian*, Jan 20th. Available at https://www.theguardian.com/uk/2011/jan/19/undercover-police-officer-lynn-watson

Vidal and Pidd (2007). "Police to Use Terror Laws on Heathrow Climate Protesters", *The Guardian*, August 11th. Available at https://www.theguardian.com/uk/2007/aug/11/ukcrime.greenpolitics

· 5 ·

"DON'T GLUE YOURSELF TO THE PRIME MINISTER!"—MILLENNIAL MEDIA MOVEMENTS AND ALTERNATIVE ACTIVISTS' COMMUNICATION STRATEGIES

> Don't superglue yourself to the Prime Minister! She said. But, I came down to London, met the Plane Stupid people and we agree it is a good way to get the message out.
>
> —Interview with Dan Glass (September 2011)

The new millennium brought with it new media and a plethora of social media platforms. New media platforms meant for ecoActivist new ways to reengage with a technologically driven media landscape with fewer gatekeepers. The internet, World Wide Web, and Smartphones were new technologies generating open reflective spaces. New spaces meant that activists, as early adopters, were able to mirror the horizontal consensus decision-making systems found in activists' movements.

However, engaging with the traditional media meant running the risk of a return to the negative reporting of the global justice/ anti-capitalists protest (see Chapter Two) that led to some collectives withdrawing totally from working with traditional media. For a long time, many collectives choose the path of least resistance with a blanket ban on talking to journalists, or if a collective did issue a press release, the information was conveyed to the public incorrectly or in a negative frame. For example, Reclaim the Streets (1990) chose to impose a media blackout, as did Class War and Whitechapel

Anarchists. Yet, media blackouts and alternative media outlets were no longer the best approach to either challenge the hegemonic green governmentality of the political parties and global institutions, or generate enough public support to support egalitarian environmentalism. Engaging with the evolving media ecology would mean a U-turn on an earlier position having chosen to remove themselves or been excluded (see Chapter Two).

Activists realized without engaging with traditional media they can't challenge the dominant discourse. By the new millennium short haul aviation activists, collective Plane Stupid and the Camps for Climate Action (aka Climate Camp) began reengaging with traditional media, but this time as a media movement.

A media movement is a relationship necessity between traditional media outlets and activists (Lester, 2007). The first UK Climate Camp emerges in 2005 as a fully formed media movement. As media movements, they saw a way to open up access to camps for journalist, whilst being able to be their own gatekeepers. In defining the parameters of where in the camp journalists could go, meant the Climate Camp's media team could act as gatekeepers. As a media movement, the team were able to select journalists, embed journalists into protest camps or actions, and media manage the message. The Climate Camps as a media movement aimed to work with traditional media to influence public opinion whilst setting the baseline as a media movement.

Whilst other radical ecoActivists collectives chose to stick with a no-media strategy, others began opening up communication channels by building websites. Websites were developed and served two purposes 1) a portal for journalists to directly contact activists 2) a means of bypassing traditional media gatekeeping and therefore filtering information to journalists. As one long-term activists explains how; "2005 became the cut of point when it became less altruistic and the middle classes joined, and then you have all of the self-promotion, and using all the media out there" (interview with L.A.R.C[1] activists, MC, 2014). However, websites and social media only has a limited reach, often to people already supportive of a cause or ideologies—preaching to the choir in other words.

Millennial media movements (engaging with traditional journalists) combine working with supportive journalists with lessons learnt from earlier actions and activists. Through innovative outreach, activists have changed the working relationship between some journalists and collectives. Technology has had an impact on millennial movements as activists use webcams to record conversations with suspected undercover police officers. Activists began to use the very tactics of law that had been applied against them to challenge 'fake news'. For example, Climate camp successfully challenged national newspaper coverage of the 2007 Climate camp (Heathrow) through the Press Complaints Council (PCC).

In a mediatized society becoming a "media movement" (cf. Gitlin, 1980; Lester, 2007) is important if activists wish to grow their support and influence public opinion. As a media movement, some groups such as Plane Stupid learnt that planning for an action had to include places that guaranteed media coverage. Couldry (2000) terms these spots as media eye. Media eyes are events or places with a high concentration of politicians and/or journalists with regular media access- such as camera's mounted opposite the Whitehouse or Houses of Parliament. Other actions inverting the media eye to the advantage of the activists was the use of supergluing activists to politicians and/or each other and planning action where there is a guaranteed media presence, even to the point of invading television studios.

However, media movements can sometimes tip over from being radical political movements into pure media movements. A high concentration of social media and online self-promotion often lacks any substance or meaning for any action. In the rush to get front-page coverage, actions lacking in long-term strategies often result in an emptying out of the reasons for a protest.

Similarly, other collectives such as the Climate camp's media team successful use and implementation of "tactics rather than laws, and even of using laws themselves as tactics" (Foucault, 1991, p. 95), enabled the media team and other media movements to challenge the negative or fake newspaper reporting of action. As this chapter will show, climate camp successfully challenged a *London Evening Standard* newspapers' front page as false by lodging a complaint with the Press Complaints Commission (PCC). One Reclaim the Power and long-term activists tells us that many of these shifts in tactics into becoming a media movements arises from lessons learnt, new technology, social media, tactics of law, community outreach, manipulating the media eye and old fashioned NVDA protest in front of the media. She says that;

> in more recent years, probably from climate camp onwards, people have learnt lessons from the pasts, and with skilful engagements with the media, there's great attempts to turn around negative stories, by putting forward skilled spokespersons, skilled press officers who've got professional experience involved in the media. Having people that come across as quite ordinary. (Interview with female Reclaim the Power activist, 2014)

Acts of civil disobedience are part of a media strategies that include image events (Deluca 1999) to manipulate the media eye (Couldry, 2000). As early adopters of online media strategies, media movements benefit the most with greater capacity for movement building through their online presence. In the new millennium, capacity building and media movement models were the two key divers of the ecoActivist movement in the UK.

This chapter draws on interviews with activists involved in many of the actions set out above. In this chapter, we look at two examples of media movement strategies through testimony from those involved from the UK Climate camp and Plane Stupid. The interview data reveals that activists have learnt lessons from previous actions and movements. Conversely, social media has created issues that Gitlin identified as problematic in the 1980s. Therefore, this chapter will firstly outline the history of ENGOs and activists as a media movement. Secondly, it examines the processes and strategies of re-engaging with traditional media as a media movement; thirdly look at ways media movements have made use of tactics of law and technology, and finally examine the challenges faced in becoming a media movement. Before, we look at the two examples, the chapter traverses through history the ways ecoActivism becomes a media movement, from the early days of Greenpeace to the millennial Climate Camps.

Early Media Movements: Greenpeace

In a mediatized society protests collectives understand that a conscious lack of engagement with professional journalists is potentially detrimental to the representation of their own activism. The result has been that many movements are also "media movement[s]" (Lester, 2010, p. 110), accepting a two-way affiliation with traditional journalists making a "media-movement relationship a necessary, but uneasy one" (Lester, 2010, p. 110). Movements have had to accept that some relationship must exist between traditional journalisms and activists, even if the journalists are complicit in framing protest in negative terms. Todd Gitlin's (1980) seminal work, *The Whole World is Watching* examines the Students for a Democratic Society (SDS) linking mobilizations [that] "were shaped in part by the codes of mass media operations. It conducted its activities in a social world that recognized it, liked it, and disliked it through media images, media versions of its events and rhetoric. To some extent the movement even recognized itself through mass-mediates images" (1980, p. 4). Gitlin goes on to note that:

> As the mass media define the public significance of the movement events or, by blanking them out, actively deprive them of larger significance. Media images also become implicated in a movements self-image ... what defines a movement as good copy is often flamboyance, often the presence of a media-certified celebrity-leader, and usually a certain fit with whatever frame the newsmakers have constructed to be the story at a given time. (1980, p. 5)

Around the same time, Greenpeace (1971) were both engaging as journalists and producing journalism for the Canadian and American media. Sailing from Vancouver, Canada to the Amchitka Island, off Alaska, a group of scientists and journalists attempted to stop USA President Nixon's government from testing nuclear weapons. The mobilization was to be the inaugural event to what eventually became Greenpeace.

Leading the charge, Robert (Bob) Hunter was a journalist with the *Vancouver Sun*. Hunter inspired by media commentator Marshall McLuhan understood the importance of media coverage to define both the activists and the group. Hunter and his team, having chartered a boat, the *Phyllis McCormack*, from Vancouver (1971) they attempted to sail into a nuclear-testing zone off the coast of Alaska. Using non-violent direct action (NVDA) their aim was to sail into the testing zone to stop the tests from going ahead. The USA tested the bomb ahead of the scheduled before the *Phyllis McCormack* could reach its destination. Regardless, as a media movement the mobilization may have failed, but with a well thought-out media strategy it created a media frenzy (Hunter, 2004).

Controlling how information was release to selected media helped maximize coverage. As journalists, Hunter and his team could provide readymade copy to the press. Journalists on board were assigned newspapers to target, and help build a rapport with editors. Hunter explains that the idea was to create a rigourous communication system.

> Cummings would file his stuff with Vancouver's underground paper, *The Georgia Straight*, which would pass it on to all the underground papers in Canada, the US, and Europe via the Liberation News Service. I would pump out a daily column in the Vancouver Sun. Fineberg would get stories out to papers in Alaska. (Hunter, 2004, p. 43)

Hunter's second tactic was to gain the support of the students in Vancouver. Linking with the student movement, worked, as the *New York Times* front-page newspaper declared, "Students protest A Test", and reads as follows;

> Nearly 9,000 elementary and high school students gathered at the United States Consulate ... in peaceful protest against the planned nuclear test ... no serious

> incidents occurred at the demonstration, although the police said that a few youths tried to remove an American flag. (Reuters, 1971, p. 55)

Hunter *et al.* (2004) set the tone for future media movement strategies and a realization that there is a need to engage with traditional media journalists. Today, Greenpeace have a highly successful media strategy that combines ENGO system with NVDA, 24-hour newsrooms and aside from the Brent Spar incident (see Chapter Two) continues to make the most of mediatized society.

Millennial Media Movements: A Balance between Openness and Protectionism

Since the end of the twentieth century, radical protest collectives have begun reengaging as media movements despite experience making many cautious about working with journalists. History has taught activists to be wary of engaging directly to the point that some groups choose to abstain from media engagement. The Whitechapel Anarchist and Class War chose to abstain and uphold a media blackout during the 1990s and into the Climate camps (2005–2010). Tents marked with "no media" signs signalled to journalists and the camps media team the area was out of bounds.

Activist working groups and individuals develop relationships with specific journalist by either embedding them in action or feeding select journalists interviews and stories. At the same time former and current journalists were joining movements, bring with them knowledge and contacts of media organizations. With former and present journalists, working alongside national newspaper journalist and co-ordinating media strategies during protest provided a stronger media strategy although heavily dependent on social media and a media movement begins to appear.

Stalwart activists MC (2011) notes, some of the earlier media coverage of Reclaim the Street (and the anti-roads protests was belittling to the activists)

> takes Swampy, where they [video journalists] shot down, which they never do to Tony Blair, who is about the same height. They shot down so you see this little guy looking up, telling you. This is what we're gonna do. And Swampy just got side ambushed" (interview with L.A.R.C[2] Activists, MC, 2014).

Another activist explains that people are wary of talking to journalists for fear of the interviews focusing on violence (as established in Chapter Two).

Stalwart Des from the Critical Mass collective reiterates the same issue, as he says,

> a lot of the time you get eclipsed by the violence, it's just a shame really I mean I think the media's got a lot to answer for really in the sense of, it's very strange isn't it because they think all this the doom and gloom sells newspapers and it just perpetuates really. (Interview with Des Kay, 2014)

By forming symbiotic relationship with specific publications (like *The Guardian* newspaper) and journalists, there was always an awareness of keeping in mind a need to adapt practices "to reflect social movement traditions, particularly in regard to critiques of representation and corporate-and government-controlled media power" (Feigenbaum, Frenzel, and McCurdy, 2013, p. 74). Indeed, a recognition that "complete or total abstention is not really possible in an age of 24-hour news cycles and social media", it is more a case of how best to garner media coverage in a "period of media saturation, as it becomes increasingly difficult for protesters to find ways to be both visible and invisible to the media" (Feigenbaum *et al.*, 2013, p. 75). Ultimately it was mainly a realization that a "self-imposed isolation ... was a luxury that we [activists] could not collectively afford" (CSC, 2005, p. 322).

Australian activists Get Up's Fight for the Reef campaign found the importance of having relationships with journalists. "It is absolutely critical (...) it allows you to get stories up, find out information, and shape the stories (...) yeah it is a critical part of what we do" (Interview with A. Dodds from Get up! Australia, June 2015). Dodd's adds that "during the fight for the reef campaign, I've worked—I've been really happy with how I have worked with the Graham Lloyd, the environment editor for the Australian. He doesn't believe in climate change as far as I am aware and he does think that wind turbines do make you sick, however he has been quite amenable to writing good stories for us around the industrialization of the reef coastline. Without having a relationship with Graham Lloyd and talking to him about that issue, we never would have got those stories up" (interview with A. Dodds from Get Up! Australia, June 2015).

Queensland conservationist, Wendy Tubman explains the importance of keeping local/ regional journalists on side "I am not blowing my own trumpet—I am just saying that I am not going to bullshit them. I am not going to make up figures, I am not going to make up arguments, and they know that they can rely on me. I am not going to get them looking stupid or anything else. They can trust what I say. If I don't know about something I will tell them" (Interview with activists Wendy, July 2015).

Organizing into a Media Movement

A media movement's communication and coordination strategies are organized in two ways, local and working groups. Local groups (5–50 people) do local media work, direct mobilization, organizes camp neighbourhood, such as communal kitchens, toilets, and send representatives to nationwide monthly meetings. The working groups (2–25 people) tasks range from running of a site, finance and fund-raising, media, networking and website; meetings and facilitation, and workshops and entertainment. Local and working groups feed into monthly national meetings (50–100 people) ahead of any mobilizations.

At each gatherings a list of previous decision, an agenda and a minute taker and minutes are posted to a website. Reclaim the Power use the website Basecamp whilst other makes use of listserv and email lists at the Riseup [dot]. net. Communication amongst the groups is done via emails, and each protest camp has its own forums (climatecamp.org), and decisions whether to include political parties, NGO or any other presence is reached by consensus and the information disseminated via the media working group and listservs. In combining previously knowledge from stalwarts and millennial savvy technologies and skills in social media (Facebook circa. 2004; Twitter circa. 2006) website coding and development, social media provides appropriate tools to have some influence over the media message.

Millennial Media Movements: Turning the Tables with Technology

Making the most of technology makes a movement innovative when engaging with journalists. Challenging journalistic practices through network power and citizen journalism is one approach, but ecoActivists have taken social media and technological advances also give innovative ways to challenge the state. A common tactic of organization is the Swarm, an old technique making use of new technology. A swarm is similar to the roosting swarms of swallows where a movement, moves en-masse, each "moves as one, as if it's one organism. Yet no-one is in charge, it seems to happen as if magically" (Notes from Nowhere, 2003, p. 67). Swoop and swarm are terms coined by the RAND Corporation (2002), and adapted by activists. In the pre-internet days organized through word of mouth Swarms would be coordinated through a maze of symbols, colour coding and whistles. A swarm during the Carnival against Capital (on 18 June 1999), 8,000 face-masks were handed out to activists. Masks were red, blue, green or black, and on a given signal (in this case a whistle) each colour would follow one person with a correspondingly coloured flag (Tyler, 2003), and would gather and swarm on a pre-decided site.

Later, with the internet, swarms were organized via text and SMS messaging. Interested people could sign up to a text messaging service, just as Zapatista supporters signed up to listserv, and they then would be texted a series of meeting places, before the final protest camp site is revealed. Technology makes the organizing of a swarm or swoop much easier, as SMS messaging means that activists can arrange the events through tweets and Smartphone messaging (Newlands, 2012). Using mobile phones and a social networking site, it opened up the event, enabling anyone with an interest (including journalists and the police) to be part of the swoop, either on a physical or virtual level.

The right image provides a ready-made narrative was also important for an action during the 2014 G20 in Australia. Environmental entrepreneur's, Eden and Bondi resident Rex, attached a camera to a drone, and flew it over an action on Bondi Beach, Sydney, New South Wales. The action, Heads in the Sand was a metaphorical image of the G20 approach to climate change. Four hundred people put their heads in the sand; the image drove the media coverage. As organizer, Eden explains, "it was all about the image. Sure, if people have a nice time on the day, and say something to the media, but really, what we were aiming for, it was all about the image. What will go around the world is that image" (interview with activist Eden, November 2014). Focusing on an image was intended for two reasons, to shift perception of protest and bring a unique news angle, "to get away from the protest word as well. When we were dealing with the cops, I kept saying it was an attempt at public art, and I believe that" (interview with activist Eden, November 2014). Using drones and art for action provides a different narrative to the usual protest story. Eden describes why they chose artivism over traditional marches, is because a march is just, "a bunch of people who all agree with each other shouting out things that they all agree with to each other, and just shouting it out to the street, and I just don't know what the real outcome of that is" (interview with Eden, November 2014). The march is formulaic and ultimately producing stock photographs with "a couple of photos of peaceful protest that look like any other peaceful protest. Those photos aren't going to get any traction in anything really, because it just looks like another protest" (interview with activist Eden, November 2014). Using a drone to capture an image in an innovative way of an old form of protest in a public display. The Heads in the Sand Image at Bondi Beach and an earlier version in Townsville, north Queensland went "viral" on click bait websites, Mashable, Buzzfeed and picked up by traditional media online sites. In preparing a ready-made narrative with a single image, using technology of drones and a handheld tablet PC to frame the image shows how social media and technology can be a good source to contribute to public understanding.

Technology and media movements' strategies came together when activists and a journalist from the *Guardian newspaper* were observing a covert recording of police officers attempts to recruit activists as an intelligence source. Plane Stupid activists, Tilly having being approached by police to become an police informant used the opportunity and technology to record the conversations. Listening in on Tilly's conversations with the police was Guardian journalist Paul Lewis and fellow Plane Stupid activist Dan. Dan explains what happens, "Tilly was leading the police on, [the spycam] was attached to Skype and I was sitting in one of the bedrooms recording it all with Paul Lewis from the Guardian" (interview with activist Dan, 2014). The innovative reconstituting of police practices through technological tools meant activists could turn a tablet or iPhone into a recording device. Just as Kennedy had used police recording devices disguised as a watch (see Chapter Four), activists equally adapt the technology available to them. Working with the journalists meant the story was "front page of *The Guardian* when Tilly was at the centre of the story, using a spycam to expose the police, that was brilliant, the spycam": (interview with DG, 2014). Lewis and Evans (2013) account of undercover policing says of the episode that;

> It is a murky world, but one that was laid bare thanks to the cunning of an environmental activist. Tilly Gifford was a member of Plane Stupid, an anti-airport expansion collective that occupied runways. In 2009, she was approached by two plain-clothes detectives in Glasgow and asked to inform on her friends. (Lewis and Evans, 2013, p. 208)

Lewis and Evans describe the sting as

> The 24-year-old decided to play the cops at their own game. She would pretend she was open to selling information about Plane Stupid, but secretly record the

discussions with the handlers. She sewed a secret recording device into her tweed jacket. Gifford ended up recording three conversations with two plain-clothes cops from Strathclyde police, a detective constable, and his assistant." (Lewis and Evans, 2013, p. 208)

In these examples, the activists had to trust the journalists to report the story fairly without prejudice; in turn, the journalists got a front-page story. Building relationships and trust with journalists is important, but so too are strategies that bring the story to the media. In a 24/7, media world, with traditional and social media vying for attention, giving a story to a journalist cuts out all the legwork. This is not a new idea, but previous attempts have caused tensions within collectives, often misinterpreted or reframed by the traditional media as fluffy or non-political. To overcome such misunderstandings, activists have devised a series of strategies learnt from the past.

Millennial Media Movements: Camps for Climate Action (Climate Camps)

In 2005, two months before the Group of eight (G8) summit, at Gleneagles in Scotland a media group that had grown out of the Dissent! Network into the Counterspin Collective (CSC), had begun shifting towards being a media movement. Dissent! Network behind the initial media team acknowledge media are "part of the problem and not part of the solution" (CSC, 2005, p. 322). There was a realization that it would be a "mistake to reject the possibility of strategically using the mainstream media outlets to promote our ideas and tackle head on the discourses of politicians, corporations/recipients of such media coverage to think differently outside their own comfort zone" (CSC, 2005, p. 322). CSC key job was the facilitation of "media relations … like a sort of dating agency for journalists and activists (…) offering a network of translators so that press releases could be distributed in many languages" (CSC, 2005, p. 324).

The G8 meeting was deliberately chosen to hold a media event given it was "viewed as part of an ongoing series of international mobilizations which have been on the mainstream media radar since 1999" (McCurdy, 2009, p. 44). The UK was swamped with domestic and international journalists with both the Group of Eight meeting in Scotland, and the second Live Aid concert happening in London, activists set up a HoriZone climate camp.

Greeting journalists wanting to visit the camp was a media gazebo outside the G8 Hori-Zone camp. Inside the camp, the media production team were producing citizen journalism via smartphone and tablets within an independent media centre. The external media tent formed a press centre where journalists were able to visit the camp with an activists escort. Over the five-year life of the movement, the tent moved into the confines of the camp (McCurdy, 2009). The effect of placing activists inside, and traditional media outside was akin to "mainstream media front stage, radical allowed backstage" (McCurdy, 2010, p. 48).

At the Blackheath Climate camp (2009) there were further restrictions on access for traditional media outlets. A camp the media team communique advises that "TV crews and press photographers are allowed in the camp between 10 a.m. and 7 p.m., so long as they have a friendly guide from the media team" (Media Team advisory, 2009, p. 12). Media spots outside of these hours were possible providing journalists "are accompanied on and off the site" (Media Team Advisory, 2009, p. 12).

A tactical media approach by CSC was to create an "open hour" (CSC, 2005, p. 324). Journalists invited onto the site were asked to sign a code of conduct. The Climate camp media team and independent media activists VisionOnTV ran workshops to be able to "do media work, from making an online sensation with your mobile phone to staying on message" (Anon, 2008, p. 9). The workshops helped to encourage and teach activists to produce their own news and commentary footage, and a series of leaflets set out how to record a one-minute news items. Lessons in using smartphones to adapt the journalistic baseline of the inverted pyramid form were shared to maximize the chances of content being aired on the mainstream media.

The strategy initially worked with CCA growing to the point that ENGO's wanted to be part of the protest camp. Historically, professional environmental activists groups and similar organizations have seen ecoActivism as the radical flank of environmental discourse. Facilitator Steve explains that,

> everybody wanted to get involved, including the mainstream NGOs some quite clumsily, like Greenpeace and some not quite as clumsily, like Friends of the Earth, and it culminated in, when we actually got to the camp, it was a media scrum, you know, the whole street was full of media and we had volunteers from NGOs in their personal capacities who did media for those NGOs who came and helped us. (Interview with facilitator Steve, 2011)

Initially, there were plans to invite journalists to be embedded within the camp, but having previously intended to invite a small pool of journalists to live on site for two days in order to facilitate in-depth coverage of the camp, climate camp had to abandon the plan for Heathrow because, it has a member of media team explains at being untenable for several reasons, that

> The camp has become a big media story and we have been overwhelmed by media interest. This has included many requests for extra or different forms of access, which we cannot meet. The inadvertent creation of such a multi-tiered access system led to criticism that we were attempting to control media coverage, which was not the policy's intention. We have not been able to find enough camp neighbourhoods willing to host journalists living on site, which was one of the pre-conditions for the longer access proposal. (Interview with media team member, July 2015)

The challenge in controlling a message is also ensuring the message doesn't get lost in the media hype. As a media movement an objective is to create a form of resistance, to challenge the power of conducting through their own forms of resistance (non-violent direct mobilization, civil disobedience, site of contestation) that offers journalists and participants an alternative' way to be conducted. CCA's media movement strategy become problematic because instead of using their own media as a form of resistance to traditional journalism in an attempts for equilibrium of power, the media movements generated power struggles within the camp, that lead to dislocation and manipulation of the message. Adopting the same journalistic tactics of press areas, media releases, passes, they were conforming to hegemonic practices, where engaging in more in a revolt of conducts is not "merely dealing with the same phenomena in reverse, from the negative or reactive side" (Foucault, 2007, p. 195), then more longevity and understanding may have been beneficial.

Media Movements: Utilizing the Media Eye

EcoActivists collective, Plane Stupid are an aviation collective against airport expansion and short-haul flights and developed one of the strongest media movement relationships to promote their ideas. Plane Stupid activists JS (2011) explains the strategy, as "using the traditional media but we've also been using the non-traditional media" (interview with HACAN Chair, John Stewart, 2011). As a media movement, Plane Stupid and other movements built a symbiotic relationship between individual journalists and later editors to publish stories that were sympathetic to the cause. Plane Stupid utilized the media eye of Parliament to stage a protest against the proposed third runway at Heathrow Airport (February 2008). Activists climbed onto the roof of the

Palace of Westminster to unfurl a banner proclaiming that Parliament was "BAA Headquarters", while a second banner carried the web address for Plane Stupid. Purposefully dressing in tweeds to represent what they felt were the gentry of Parliament Plane Stupid activists once scaled onto the roof then handcuffed themselves to railings. As Plane Stupid activist and HACAN chairman, John Stewart, notes:

> when the five or six people who went onto the roof of The House of Commons, they were very clear in the early meetings that we didn't want headlines like that [points to *London Evening Standard* language of militant to describe activists] so even to the extent of what they would wear. A decision was taken that they would dress up like the sons and daughters of *Daily Telegraph* readers to attract those papers … it so happened that the Daily Mail actually came to them and wanted to do a feature. (Interview with, John Stewart, 2011)

Plane Stupid's approach was "to try and use the media by subverting it, by using their own tools" (interview with John Stewart, 2011). In deliberately attempting to subvert the media by taking protest to the media eye, they were almost guaranteed coverage from the many journalists and permanent television cameras around Parliament Square. The mobilization achieved its goal, and Plane Stupid had front-page coverage in all the London newspapers. Social media meant those on "on the roof of the House of Commons, [the Plane Stupid people] were taking the pictures. Those pictures were going to Sky, BBC before anybody else could, so Plane Stupid set the agenda, their version of things and I think it helped in the immediacy of organizing" (interview with activist JS, September, 2011).

However, even aiming for the coveted front-page, they were not able to control the language and it reverted to negativity and security discourse. London regional newspapers. *London Lite (since closed)*, the *London Evening Standard* and the *London Paper* (since closed) all ran with an image of the protesters on the roof. Headlines announced, "Airport protesters make a mockery of Commons security" (Murphy, 2008), and The London Paper called the protest the "Storming of Parliament" and "Security alert at Commons: protesters scale Parliament" (Sutherland, 2008).

Media Movements: Superglue—Stick it to the Man

Media eye manipulation occurred again when Plane Stupid's activists Dan Glass was the recipient of a Sheila McKechnie Award Foundation (2008).

The Sheila McKechnie Foundation rewards those campaigning against social injustice, in memory of Dame Sheila McKechnie,—a Scottish Trade Unionist, and housing advocate. Dan was being honoured for his work in raising awareness of the health impacts for communities living under flight paths. He was invited to receive his award from the Sheila McKechnie Award Foundation Patron British Prime Minister, Rt Hon. Gordon Brown at Ten Downing Street. As a media movement, the opportunity presented a perfect opportunity to subvert the media eye and make a statement. A plan was hatched that Dan would superglue himself to the PM. Dan explains that;

> We wanted to highlight the hypocrisy of the award. I chatted to a friend before, asking them what we can do to highlight this. Why don't I superglue myself to him?" Just as a joke initially, and she was like, "Don't superglue yourself to the Prime Minister!" "Ohhh, it's a really good idea." I came down to London, met JS, and other Plane Stupiders its like, "I think this is a good way to get the message out there", and that was that really. (interview with Dan in September 2014)

Like the action on the roof of Parliament meticulous planning went into the superglue plan to ensure the most media coverage. Discussions were had on what Dan would wear, how and where he would superglue himself, too the Prime Minister, what would be said to the media, and which media might be given a tip off. Plane Stupid activists John (2011) explains how it happened:

> I think one of the reasons is that Plane Stupid were thinking really hard beforehand, you know, what would Dan look like? Would he wear the glasses that had no lenses in them just to look a little more respectable … and it worked.

Prior decisions were made, whether or not, "Dan would superglue himself to his hand or to his jacket?", and there was concern that "the hand in some ways is more dramatic as that could [seen as] be assault". It was decided with the jacket the Prime Minster could just remove that jacket, and not be stuck to the activists. John adds "It was a wonderful stunt, wonderfully cheeky. But we had to sell it in that spirit of a cheeky but responsible stunt".

On the day, Plane Stupid activists Dan stands in front of UK Prime Minister, Rt Hon Gordon Brown M.P's (2008). Dan calmly explains to the Prime Minister that he has superglued his left hand to the Prime Ministers right sleeve in protest over the contribution short haul flights make to climate change. Once Dan feels the media have taken the appropriate photographs and the press have their story, he tells the PM he is free to remove his jacket at any time. The turning to the assembled journalists and guests in an agreeded script, "Do not worry, this is a non-violent protest. We cannot shake away climate change like you can just shake away my arm" (Interview with Dan, June 2011). The entire event was a calculated piece of non-violent direct action seeking to gain the most out of subverting the media eye. However, not all Plane Stupid's felt the coverage was positive, hey did try to make it quite silly though. At the end, they kind of, they are like, "Oh, but he's just a bit of an idiot, though." (Interview with activists NR, September 2011)

As a media movement attempting to overcome the negative representation through such actions was, for Dan a success. As Dan notes, "this is actually really useful for the powers that be who try and make me out to be an extremist," adding that if the PM's remobilization was different, the consequences for Plane Stupid could have been more dramatic as the earlier discussion in Chapter Two shows. Ultimately, Plane Stupid felt the mobilization was taken with good humour, and even "but the Head of State thinks it's alright". (Interview with Dan, September 2011). Dan later told the press "The PM laughed and promptly moved on ... He was just grinning about it. He didn't seem to take me seriously-said Glass, who was allowed to stay at the event for a further 40 minutes before trying to glue himself to the gates of Downing Street" (Summers, 2008).

Superglue remains a tool to gain media attention; anti-fracking activists have superglued themselves to the doors of the Department of Environment, Food, and Rural Affairs (Defra), and energy company iGas, the UK's largest shale gas company (2014). The mobilizations were part of a day of mobilization by Reclaim the Power's camp in Blackpool, North West England. The Camp highlighted the plans for coal seam gas with mobilization across Lancashire County. The superglue and Houses of Parliament actions were built on lessons learnt from the Swampy and Greenham Common Peace Camp reports.

Millennial Movements: Adapting from Lessons Learnt

At the Greenham Common Peace Camp (1982–2000) the media slogan "all the stars are in the sky" was a mantra for all to say no one person can speak on behalf of everyone. This mantra has become embedded in many media movements. Plane Stupid, Climate Camp, Reclaim the Power and many other collectives have taken similar lessons from the past. Lesson learnt from previous movements and action, in particular Swampy, today underpins media movement strategies, avoid what Gitlin (1980) identifies as "certified celebrity leader" by rotating activists in and out of the media spotlight when action take place in the media eye. To avoid all the attention on one person, and that person becoming the poster girl or boy for a collective, they the climate camp media team, and other collectives such as HACAN and Plane Stupid, learnt that a focus on one activist could become problematic for the whole movement. John from HACAN explains the reasoning for the decision as being,

> One of the things was, if someone was in the media for a while, we accepted that for a they would go back … having been through the roads protest thing where the media did mess us up, kind of, big time, is, we had a very clear strategy from the very beginning how to do it, so we tried to be in control of what we were doing and without that strategy, of course we messed up from time to time, but without that strategy and that clear kind of vision it would have been quite difficult. (John Stewart, 9 August 2011)

Away from the media eye, as a mode of challenging the divide and rule reporting around the Heathrow camp, was a successful strategy known as Adopt-a-Resident—scheme between the local communities of Simpson and Harmsworth and the protest camp. Sipson and Harmsworth known as the Heathrow villages (pop. 13,750 approx.)[4] will be demolished to make way for the third runway, if approval is given.

Countering Media Reporting Attempts to Divide and Rule

Ahead of the 2007 Climate Camp, the London *Evening Standard* newspaper's front page declares that "Extremist to hijack climate camp demo: Militants in plot to paralyse Heathrow," (13th August 2007, p. 1) (Mendick and Rosser, 2007). The headline was part of larger campaign to discredit and cause divisions inside and outside the camp. As activist Steve notes, "The Evening Standard set out to deliberately discredit the camp, to find militants that just were not there" (interview with activists Steve, September 2014). By creating a sense of fear and the unknown, was as one activist from the media team says, "I think what they wanted to try and do with this sort of headline is say to young and older people who might not have been that involved but who might have thought,oh I might go to Climate Camp I like the sound of that. Oh my God!" (Interview with activists Steve, September 2014).

The *Evening Standard's* headline was for some the peak in a sustained campaign against the camp. HACAN chair JS says "a month before this came out there were a couple of character assassinations, what we would call them, of various people in the movement" (interview with JS, September 2014).

As well as character assassinations, there was also a feeling that journalists were attempting to cause divisions between the local Heathrow community villages of Sipson and Harmondsworth. As ecoActivsts Mick says, "the *Evening Standard* was very interesting here, because the *Evening Standard* was

LONDON'S QUALITY NEWSPAPER

Evening Standard

Monday 13.08.07 Incorporating The Evening News www.thisislondon.co.uk 50p

Babylon Bounce

Beach Babylon author on her private torment and triumph PAGES 30 & 31

SHARES RALLY PAGE 26

Extremists to hijack climate change demo

MILITANTS IN PLOT TO PARALYSE HEATHROW

- Hoax packages to cause security alerts
- Assaults on airport fence and shops

MILITANTS at a Heathrow protest camp are plotting to bring the airport to a halt, the Standard reveals today.

Climate change activists plan to use illegal tactics such as hoax suspicious packages to cause maximum disruption at one of the busiest times of the year.

They have also discussed simultaneous assaults on the airport's security fence to stretch police resources to the limit. Some plan to pose as customers to get into McDonald's and Star-

EXCLUSIVE

ROBERT MENDICK
Chief Reporter

bucks in the terminals and then cause trouble. More than 3,000 campaigners are expected to gather near the airport in a week-long protest against runway expansion plans.

Organisers say demonstrators simply want to draw attention to climate change and do not intend to disrupt people's holidays. But hardcore protesters, some veterans of clashes at G8 summits of world leaders and Reclaim The Streets campaigns, plan to cause maximum disruption to make the public "sit up and take notice".

Airport bosses have stepped up security around the huge site. A total of 1,800 police from three forces are on patrol and airlines have hired additional security men to guard their offices and warehouses.

Police and airport operator BAA were outflanked when a group of 150 protesters, made up of organisers and "inner sanctum" members, set

Continued on Page 4

No, no, no: Amy Winehouse storms off after a shouting match with her father over her latest drink and drug binge **Full story: Page 11**

Amy's on mend says her father

Weather 2 Comment 12-13 Londoner's Diary 14-15 Business 26-29 Life&Culture 30-37 TV 38-40 Entertainment 41-44 Stars 44 Readers' Views 45 Games 46 Sport 53-60

Figure 5.1. London Evening Standard front page, militants in plot to paralyse Heathrow. Image courtesy of the London Evening Standard.

trying to divide and rule … what they were trying to do here, was say, people coming to the Climate Camp, all those militants, those dreadful people, they're not like the people we talked about three days ago" (interview with Mick, August 2014). In addition, Plane Stupid activist Dan adds that:

> what they [state & media] wanted to do was create divisions between activists and workers as well, because of not just the strategy that had gone on between, the community engagement strategy between the residents and activists but also workers and activists as well helped, you know, counteract this. (Interview with Dan, June 2015)

Another stalwart activists JJ notes, "I think the state doesn't like to see movements happening as twin strands of DNA, and is constantly trying to split them. That's happened successfully in the 60s, and resisters got burnt out. The split happened in the late 1960s and 1970s, some into politics and others into the military and internet research developer tools" (interview with JJ, May 2012)

The reasons for trying to cause division was the using a tactic of law to stop the camp by the government and British Airport Authority. As JS says, the headline "reflects the real concern they [BAA/government] had about the climate camp at Heathrow. That the airport authorities, the government, and the police had, I think they wanted to try to discredit it beforehand. I think they also wanted to make sure as few people as possible came to the camp beforehand" (2013). JS adding that

> I think it was at the same time as they took two people and myself from Plane Stupid, and we were served an injunction not to go to the camp. They also injunction [sic] the entire organisations in Airport Watch and H.A.C.A.N. We challenged that and most of them weren't injected except myself and two guys from Plane Stupid but it was part of the same thing to try and say, this is going to be dreadful; this is going to be full of Militants. (Interview with JS, September 2013)

Whether the headline has any truth or not is a matter for discussion, but as activists MC (2013) says, "Climate Camp is a very open movement and probably somebody within Climate Camp, whether it was a journalist come to a meeting, probably actually said these things or an undercover policeman, whatever, they probably were said. But, a journalist's job is actually to find the truth not just something that fits their opinion and publish it". Whilst another does not agree, believing the headline is a clear media strategy to create tensions and cause splits in the movement. Activist Steve says that (2013),

> I think it is very unlikely that somebody did say those things; and it is no surprise that it is in the Standard. Obviously smear tactics to stop people from coming but I think, politically, importantly, it's to try to distance us, for want of a better description, from the local residents, to convince the local residents that we were the enemy, the Black Bloc; we were going to come in, smash up stuff, cut the fences down, run on the runway. All those sorts of things were to try to get division between the local residents and us and it failed absolutely apocalyptically because of the strategy of the police of stopping and searching local residents.

Media attempts to divide and rule is not unique to the climate camps. In Australia News Corp International, publications have been framing the argument over the impacts of climate change and the decline in health of the Great Barrier Reef as

> Opponents in the mining industry and their friends in the *Courier Mail* tried to run a different sort of set of framing and positioning around that. They tried to split up the environment movement from regular Queenslanders or Queensland punters. They are trying to tell us how we can lose our things or how they are trying to restrict our economic benefits—usually the way they do it is they are trying to restrict jobs and prosperity. (Interview with Get Up! Adrian Dodds, October 2015)

However, whether supportive journalists or embedded journalism into a movements, unlike previous movements who failed to publicly change such headlines, this time the Climate camp media team and others decided to go to the Press Complaints Commission (PCC). Alex Harvey (a pseudonym) on behalf of the camp stated in his complaint that "there was nothing in the headline to indicate to readers the insubstantial basis of the claims" and furthermore that

> neither was there adequate qualification in the text of the article. And within this context, there is a failure to make clear the limited numbers of those who allegedly planned "to pose as customers to get into McDonald's and Starbucks in the terminals and then cause trouble" was also misleading.

The Press Complaints Commission (PCC) found in favour of the Climate Camp media team, ruling against the *London Evening Standard*. The two Evening Standard found journalists Mendick and Rosser[5] had no basis for either the headlines or the stories. The ruling was based on the grounds that Mendick's story was based on the journalist overhearing a group of four people when "a man in his late 20s" said "We need to make people sit up and take notice. Leave some packages around Heathrow. That'll make them take notice" (PCC, 2008). This quote, from an unnamed source, was the basis for the front-page headline.

Following the PCC ruling against the *Evening Standard* the Climate Camp media team went back to the PCC over coverage by *The Sun*, *Daily Telegraph*, *Daily Express*, and *Daily Mail* newspapers. This time their claim was over their failure to ensure clarity that allegations raised by the *Evening Standard* were unfounded, and therefore a breach of the PCC Code. The matter was resolved with the following statement from the *Evening Standard*, "further to the article first published on this website on 14 August 2007 [sic] regarding 2007 Camp for Climate Action at Heathrow Airport, we are happy to make clear we had no evidence anything other than non-violent civil disobedience was being planned".[6]

A second case was brought to the PCC attention by media activists over online reporting of the G20 Camp for Climate for Action by the *Daily Mail* (18th March 2009) and *Evening Standard* news websites (Mendick and Rosser, 20th March 2009). The complaint was that the camp was the only group named in the forthcoming G20 mobilization under the headline "Anarchists plan City riots for G20 leaders' arrival". The complaint was that the "headline had referred misleadingly to 'riots' being planned by anarchists. The group was planning a blockade and not a riot". A resolution was reached and a change to the online headline "riot" to "demo". The following statement published by the *Daily Mail* reads

> Further to our article of 20 February "Anarchists plan City riots for G20 leaders' arrival" we are happy to clarify that Camp for Climate Mobilization were planning a demonstration and not a riot to coincide with the arrival of world leaders for the G20 summit. We apologies for any misunderstanding caused by our headline.[7]

The *Daily Telegraph* printed the following clarification "The Camp for Climate Action would like to point out that they had no plans to plant hoax bombs or assault the security fence" (Millward, 2007), and amended the online headline to "Heathrow protesters may stage bomb hoax". The Climate Camp had some success in challenging the negative coverage, but activists realize that there needs to be more than just being a media movement.

Promoting the ideas of the movement was another tactic for the Plane Stupid collective, aiming their media stunts at subverting the media to promote ideas within a media movement. Beyond the Heathrow villages, activists also draw on the wider community, in particular targeting journalists and politicians living within a 15-mile radius of Heathrow of West and South West London. HACAN spokesperson, John says, "a lot of the Evening Standard readers and journalists lived under the flight path to Heathrow, and they were against a third runway at Heathrow, they couldn't do anything else. It turned out towards the end of the campaign really quite useful as a traditional media in our campaign and we used that".

Heathrow climate camp was media strategy was too highlighting close connections between business and governments, through the BAA injunction (see Chapter Four) making strong links with the community to reclaim and devalue the term militant, and to work alongside NGOs, all strengthened the campaign serving to increase, not decrease public awareness. As JS explains by working with "local campaigners in an organization which represented the people whose homes were to be knocked down, they were struggling beforehand, actually did a lot for the Climate Camp but it also did a huge amount for the Heathrow campaign because it became global after this. It became, certainly a national issue".

Although many activists see the Heathrow campaign as a watershed moment for the movement, they are also aware it created long seated divisions. Activists NR says, "having done a lot of slagging off of the Camp process, is that, I think overall it was a success because the main original objectives of the camp was to build the movement to the extent that the neighbourhoods could go off, organize their own camps, have the capacity and wherewithal to do that if they so wanted to, and we did, so it was a success" (interview with NR, 2013).

"Slagging off" became a problem toward the end of the Heathrow camp, and into the following camps at Kingsnorth (2008) and Blackheath (2009). Inevitably, just as Todd Gitlin had observed in his study SDS, the climate camps had begun to see themselves through the lens of a media perception. As Gitlin notes that,

> As the mass media define the public significance of the movement events or, by blanking them out, actively deprive them of larger significance. Media images also become implicated in a movements self-image (…) what defines a movement as good copy is often flamboyance, often the presence of a media-certified celebrity-leader, and usually a certain fit with whatever frame the newsmakers have constructed to be the story at a given time. (1980, pp. 4–5)

Just as Gitlin found, so too problems began arising when the climate camps attempted to gatekeeper the sites to the extent they believed the self-images being shone back to them through the media. Climate camp had fallen into the same trap that countless other media movements had, found it impossible "to ignore the media spotlight or to turn it at will to the movement's own uses" (Gitlin, 1980, p. 5).

Traditional Journalism Reject the Media Movement

Over the years, the media movement strategy saw journalists become more and more resentful at having to be escorted, and only given access during media spots. Frustrated journalists hit back at the camps gatekeeping techniques such as signing a code of conduct. Journalists felt ostracized and excluded from the camp without legitimate reasoning.

In an open letter to the media team at the Kingsnorth Climate Camp, the National Union of Journalists (NUJ), Freelance Organizers, John Toner wrote to them asking to reconsider their open hour and access policy. The letter reads:

> *Dear Media Team,*
>
> *Last year I wrote to you raising our concerns over your restrictions on media access to your Climate Camp at Heathrow Airport. I am writing to you again, at the request of many of our members, to protest at your arrangements for this year's Camp at Kingsnorth. In a similar way to last year's procedures, you intend to restrict the media to "facilitated tours" of the camp between 11 a.m. and 1 p.m. To limit access in this way severely hampers the ability of the media to provide objective and accurate coverage. You then go on to state that journalists will be permitted to take part in the camp outside the two-hour tour, provided they sign a code of conduct. You might be aware that journalists are already governed by codes of conduct, including that of the NUJ, which you can read here. http://www.nuj.org.uk/innerPagenuj.html?docid=174*
>
> *We consider it unacceptable that you should seek to control journalists in this way. You also intend to issue a "press badge" showing name and affiliation. You should be aware that professional journalists will be able to identify themselves by producing their National Press Card. Many freelance journalists will not have any affiliation, as they are not working on behalf of any particular media company. As an organisation that espouses openness and transparency, you risk accusations of hypocrisy if you follow such procedures restricting the media. It is not too late to change this. Would you please reconsider your media access policy while there is still time?*[8]

The media team responded by pointing out "the climate camp seems to be measured against a standard of access that is rarely found in the real world. Journalists do not get to walk into people's homes or workplaces without an invitation, and when there, they are rarely given carte blanche to enter spaces considered private or restricted, for whatever reasons, nor interrupt meeting or disrupt people's work".[9] The response annoyed journalists, who reminding that media team that the camps were held in public space on common land, and therefore open to all. The media, already disgruntled, noticed the tension in the Blackheath camp (2009) and began focusing their report on class and white privilege.

The *Daily Telegraph* newspaper defines the camp as "the cheapest- and chic-est date in the summer festival calendar". The metaphor frame is repeated from the earlier reports, as protesters, are "of the 'tree-hugging variety' and 'children of the privileged" (West, 2009, p. 30). *The Daily Mail* pushes the white privilege concepts as participants are "nice, white and middle-class (…) students who have made their way to Blackheath via a summer of Glastonbury, Inter-railing, camping in the south of France" (Fryer, 2009, p. 6). Class is a common theme, as "posh upper-class white people" and the camp "looks more like a gentle middle-class festival than a political training camp" (Fryer, 2009, p. 6). In addition, as one Guardian journalist, asks is the protest camp at risk of becoming "climate camp, the institution"? (Beaumont, 2009, p. 27).

Feigenbaum *et al.* (2013) show there was an understanding that "the camp was a sanctum that shouldn't be polluted by representation" (2013, p. 69); and if the sanctum was broken by either journalists, film crew's or in one case a documentary film crew, the action caused mistrust within the camp—and "one day, as a film team did enter the camp, they were immediately surrounded by angry protesters and asked to leave. The policy was defended; the camp was not to be represented. Afterwards there were hardly any pictures or videos from the camp and this was a shame. For the most part the camp had been amazing but we didn't have many pictures to show it" (Feigenbaum *et al.*, 2013, p. 69). Splits were also widening within the camps, as Hamish says, "it wasn't long before everyone faced challenges—there was a tension within the media team, I remember we put up a banner, and someone came and took it down and replaced it with a Climate Camp banner. There were some strange things happening outside that [media] tent. And I wanted us to get on much better with the mainstream outreach team" (interview with VisionOnTv Hamish, 2011).

The following climate camp in 2010 was to be the final event. Fears and confirmation of continued infiltration and surveillance by the state, combined with in-house dislocation and the disintegrating of a previous strong media movement were all factors ending the climate camp movement in the UK. In striking similarities to the Reclaim the Street movement, a near identical repetition of police pressure, individuals talking to journalists despite a consensus agreement not too, and say the apparent quietening of the Climate camp movements.

Conclusion: Gatekeeping Beyond the Movement

Working with the media is inevitable today, but having some sense of control is important for any ecoActivists. The media movement is one way that activists can have some control of the message, but as this chapter has shown media movement comes at a cost of unity within the camps. In Chapter Seven, the space of the camp becomes an important part in retaining control. Another way is through social media. Many hail social media as the utopian solution, and whilst it does offer some freedom, an excellent tool to organize, and co-ordinate mobilizations, like media movements it has problems in how data is stored, and the idea of social capital (Putnam, 1993).

The next chapter will explore further why social media is central to media movements, and how activists use social media to develop a gap in shifting journalistic practices that has yet to be fully exploited. Regardless of limited resources, anyone can influence public opinion over an environmental concern. Democratic change maybe aided by tweets and Facebook events, but an over reliance on social media as a protest tool can also be detrimental to wider democratic practices. The next chapter shows how social media should just be seen as just one of many tools in the environmentalist's toolbox, and that too much emphasis on social media strategies and less on long-term goals, leads to dislocation and disadvantages the activists message. The following chapter concludes with a word of caution in that although social media is hailed by some as helping to foster greater democracy, nevertheless an over-reliance on social media strategies, can result in a loss of focus with too much emphasis on "hits" and "likes" and less on aims and objectives of a protest.

Notes

1. The London Action Resource Centre (LARC) is a collectively run building providing space and resources for people and groups working on self-organised, non-hierarchical projects for radical social change. https://larc.space/
2. The London Action Resource Centre (LARC) is a collectively run building providing space and resources for people and groups working on self-organised, non-hierarchical projects for radical social change. https://larc.space/
3. The London Action Resource Centre (LARC) is a collectively run building providing space and resources for people and groups working on self-organised, non-hierarchical projects for radical social change. https://larc.space/
4. http://www.hillingdon.gov.uk/article/29826/Site-search?q=Harmsworth+population

5. Several attempts were made to contact the two Evening Standard Journalists for comment for this book, but I received no response to emails or telephone calls. Robert Mendick is the Chief Reporter for the *Telegraph* newspaper in London.
6. http://www.pcc.org.uk/cases/adjudicated.html?article=NTYxMQ==&type=
7. http://www.pcc.org.uk/cases/adjudicated.html?article=Njc3NA==&type=
8. https://www.UK UK indymedia.org.uk/en/2008/07/404585.html
9. https://www.UK UK indymedia.org.uk/en/2008/07/404585.html

References

Anon. (2008). "Standard Criticised for 'alarmist' Heathrow Story", March 20th, *The Guardian*. Available at http://www.theguardian.com/media/2008/mar/19/associatednewspapers.pressandpublishing

Beaumont, P. (2009). "The Climate Camp is Too Self-regarding to be Effective", *The Guardian*, 30 August [online]. Available at http://www.guardian.co.uk/commentisfree/2009/aug/30/peter-beaumont-climate- camp-protesters (accessed 1 September 2010)

Couldry, N. (2000). *The Place of Media Power: Pilgrims and Witnesses of the Media Age*, London: Routledge.

Counter Spin Collective (CSC). (2005). "Media, Movement(s) and Public image(s): Counter spinning in Scotland", in D. Harvie, K. Milburn, B. Trott, and D. Watts (eds.), *Shut Them Down! The G8, Gleneagles 2005 and the Movements of Movements*, Leeds: Dissent and Automedia, pp. 321–331.

Daily Mail Reporter. (2009). "Anarchists Plan City Demo for Day G20 Leaders Arrive in London". Available at http://www.dailymail.co.uk/news/article-1151034/Anarchists-plan-City-demo-day-G20-leaders-arrive-London.html#ixzz4o1uxgjVn

DeLuca, K. (1999). *Image Politics: The New Rhetoric of Environmental Activism*, New York, NY: Guilford Press.

Feigenbaum, A., Frenzel, F., and McCurdy, P. (2013). *Protest Camps*, London: Zed Books.

Foucault, M. (1991). "Governmentality", in Burchell *et al.* (eds.), *The Foucault Effect*, Hemel Hempstead: Harvester Wheatsheaf, pp. 87–104.

Foucault, M. (2007). *Security, Territory, Population: Lecturers at the College De France, 1977–1978*, New York, NY: Picador, pp. 333–363.

Fryer, J. (2009). "What do we want? A nice Cath Kidston tent. When do we want it? Now!", *Daily Mail* (London edn.), 28 August, p. 22.

Gitlin, T. (1980). *The Whole World is Watching: Mass Media in the Making & Unmaking of The New Left*, Berkeley, CA: University of California Press.

Hunter, R. (2004). *The Greenpeace to Amchitka: An environmental odyssey / Robert Hunter; with photographs by Robert Keziere*. Vancouver, B.C.: Arsenal Pulp.

Lester, L. (2007). *Giving Ground: Media and Environmental Conflict in Tasmania*, Tasmania: Quintus Publishing, UTAS.

Lester, L. (2010). *Media and Environment: Conflict, Politics and the News*, Cambridge: Polity.

Lewis, P., and Evans, P. (2013). *Undercover: The True Story of Britain's Secret Police*, London: Faber and Faber.

Millward, D. (2007). "Heathrow Protesters 'May Stage Bomb Hoax'", *Telegraph Online*, Retrieved from http://www.telegraph.co.uk/news/worldnews/1560248/Heathrow-protesters-may-stage-bomb-hoax.html

McCurdy, P. (2009). *I Predict a Riot' Mediation and Political Contention: Dissent!'s Media Practices At the 2005 Gleneagles G8 Summit*. PhD thesis, London School of Economics and Political Science. Available from http://etheses.lse.ac.uk/5/ (accessed June 2012).

Mendick, R. (2007). "Extremist to Hijack Climate Camp Demo. Militants in Plot to Paralyse Heathrow", *London Evening Standard*, 13 August, p. 1.

McCurdy, P. (2010). Breaking the spiral of silence: unpacking the "media debate" within global justice movements. A case study of Dissent! and the 2005 Gleneagles G8 summit. *Interface*, 2(2), 42–67.

Murphy, J. (2008). "Airport Protesters Make Mockery of Commons Security", *London Lite*, February 27, p. 7.

Newlands, M. (2012). "Reclaiming the Media: Technology, Tactics and Subversion", in E. Cudworth, P. Senker, and K. Walker (eds.), *New Horizons: Technology, Society and Inequalities*. London: Peter Lang.

Notes from Nowhere (Organization). (2003). *We are Everywhere: The Irresistible Rise of Global Anti-capitalism*. New York, NY: Verso.

Putnam, R. (1993). *Making Democracy Work*. Princeton, NJ: Princeton University Press.

Reuters. (1971). "Students Protest A-Test", *New York Times*, 7 October, p. 55.

Summers, D. (2008). "Climate Change Protest Comes Unstuck", July 23rd, *The Guardian*. Available at http://www.theguardian.com/politics/blog/2008/jul/23/stuckonyou

Sutherland, K. (2008). "Protesters Scale Parliament", *The London Paper*, 27th February, p. 4.

Tyler, W. (2003). "Dancing at the Edge of Chaos: A Spanner in the Works of Global Capitalism", in *Notes From Nowhere: We are Everywhere: The Irresistible Rise of Global Anti-capitalism*, London: Verso, pp. 188–195.

West, E. (2009). "Climate Camp: What Tyler Would Have Felt at Home Among the 'fluffys' in Blackheath", *Daily Telegraph*, 29 August [online]. Available at http://www.telegraph.co.uk/earth/6111519/Climate-Camp-Wat-Tyler-would-have-felt-at-home-among-the-fluffys-in-Blackheath.html (accessed 1 September 2009).

· 6 ·

ACTIVISM IS MORE THAN HITS AND LIKES

Social Media Strategies and the Moveable Middle

> In terms of media, it's clear for movements in general, or all social movements. We need a massive increase in the quality and quantity of citizen media; we need to saturate the airwaves.
>
> —Interview with activist Richard Hering (September, 2011)

Social movements, especially ecoActivists and Global Justice movements are often early adopters of new and social media platforms, and had some success by combining the media movement concept with social media and online presence (circa. 2005). Social media platforms exploded onto the media stage, and activists were some of the first to understand that these new platforms could enable them to circumnavigate gatekeeping practices of the traditional media. In just three years, one of the first social media sites My Space (2003), was soon followed by Facebook and Flickr a year later (2004) and by 2006 microblogging site twitter.

With new technologies (smartphones, laptops, and handheld computers) also changed how we consume traditional news media. With more choice of where and when to consumer and produce news, smartphones were providing instant news access, with the first iPhone (2007) and Android (2008) making news portable and accessible anywhere with a 3G, 4G or Wi-Fi signal. The technology combined with new platforms meant social movements and individual activists could instantly upload news footage of mobilizations. Social media and the internet have changed the way ecoActivists engage with the general public, which has led to many collectives also becoming media movements in attempt to bypass traditional media (Hutchins and Lester, 2006). This chapter looks at the ways in which ecoActivism movements have adapted to the techniques of dominance to challenge the hegemonic discourse and reclaim a voice within the mainstream media. In doing so, they are targeting what this book is linking to the concept of the Moveable Middle. By circumventing the traditional media gatekeeping and combining the concept of a media movement with social media provides the tools to influence the moveable middle.

The moveable middle is the undecided section of the population that do not necessary what to D-lock themselves to an airport fence, or consume away the problem, but want to find do something in seeking solutions to climate change. The moveable middle is a target for both the political parties and the protesters with both Sides seeking to persuade and coerce the moveable middle in support of their principles, ideologies and ideas. The moveable middle is a term that has its origins in LGBTQI[1] (Laser, Erickson, and Kessler, 2010) movements. The advertising industry has also co-opted the moveable middle concept in their rhetoric for brand campaigns.

The moveable middle is the section in society that is undecided on their position within a discourse. The moveable middle makes up approximately 60% of a group target audience that political groups hope to influence. "There's an unwritten rule about persuasive speaking: 20 percent of audience (the choir) probably agrees with you; 20 percent of your audience might be too hard to reach; and the remaining 60 percent make up the 'moveable middle'" (Bowles and Nadon, 2013, p. 274). Organization in ideas about gay rights and same sex marriage, a study by the Third Way point out that "In order to reach the middle, we must not only identify and emphasize the values that we share with them, like equality and human dignity, but we must also address their concerns head on" (Laser, Erickson, and Kessler, 2010, p. 1).

In the LGBTQI example, the middle are people identified as sitting on the fence over same sex marriage, and are easier to reach than outright opponents to the idea. Those in the middle may know of someone or be aware of people who are campaigning or involved in the field. They are more likely to have shifted position towards the middle ground over the last 5–10 years, and become more open to ideas. In general, the middle "share certain values with the pro-equality base (…) these shared values make the middle more reachable than people in the anti-base, especially if advocates can tap into those values when making the case for relationship recognition measures" (Laser, Erickson, and Kessler, 2010, p. 3). The moveable middle has been applied to a number of fields, from retail to political science and political cultures and civil rights activism. Here we are using the term to explore how ecoActivists moments use the idea of media movements to persuade the moveable middle that egalitarian over economics are the best solution to addressing climate change.

People protest because of an experience that leaves them wanting to redress "such as experience of illegitimate inequality, feelings of relative deprivation, feelings of injustice, moral indignation about some state of affairs, or a suddenly imposed grievance."[2] (Stekelenburg, Roggeband, and Klandermans, 2015). Environmental grievances, concerns and inequalities can come from single issue events when an individual's town, field or community has been affected by a road building plan, a new runway, a coal seam gas plant. Environmental concerns also come from wider political and economic policies, for example, much of the environmental protests.

Not until the Internet and World Wide Web (1989) became more readily available (1996 onwards) many ENGO, protest movements and ecoActivist could activists invent ways to circumnavigate such media governance (Hutchins and Lester, 2009) by targeting the moveable middle via social media, in order to retain their voice within public discourse. The internet meant the anti-capitalists/global justice movements as early adopters, could bypass the traditional media (Hutchins and Lester, 2009) and find their voice in alternative media sites such as Indy media, *Do or Die* and *Schnews*.

Prior to the new technological developments early internet systems meant social movement often communicated through chat rooms and list serves. Listservs as electronic mailing list and information distribution system were useful to coordinate interested media, generate publicity, and organize actions. Social media platforms such as Facebook and Twitter meant many traditional organization and media strategies were able to reinvent themselves through social media and networks. Supporting a campaign via social media with several bespoke websites working as the face of an environmental movement or issue has become common practice today, to give the impression a movement is often larger than one or two groups. Such tools make it much easier for collectives to be media movements, give activists a voice to be able to, return as primary definer in traditional media coverage. Therefore, the internet provides an opportunity to "bypass corporate, multinational media" and to "become another device in the strategic toolbox of the environmental movement for gaining mainstream news media access" (Hutchins and Lester, 2009, pp. 580–581).

News consumers could become producers (blogs and vblogs, citizen journalism, You Tube, Instagram, and Flickr) and so too journalists become consumers (Couldry, 2003). Information becomes part of a two-way system between the journalists and the public, a horizontal network, sharing information to become "capital's dream of superfast networks that will spread consumerism across the planet" (Notes from Nowhere, 2003, p. 65). Kawamoto (2003) shows, the "internet meant journalism became an old practice in a new context—a synthesis of tradition and innovation" (2003, p. 4). With the internet, smartphones and portable news platforms, and 24 hours news cycles alters how we now consume news and our news habits.

The internet, social media, and website have more in common with the traditional media through their structural make up and organization. Traditional media tends to follow a top-down, flow of information from the one to the many. Social movements and activists mirror their consensus decision-making processes in their media strategies with a bottom-up, or network flows of information. Anyone can contribute and disseminate information. It is the horizontal, rhizomatic networks (Pickerill, 2003) of the internet that mirrors the endogenous individual responsibility model of social movements. A "rhizomatic structure provides multiple entryways, facilitating potential participants' entry into environmental protest through connections to their rhizomatic online networks" (Pickerill, 2003, p. 24). The rhizomatic nature of the Internet and new technological developments in modes of communication opens up a gap between mainstream and alternative media forms. The correlation between the "non-hierarchical" (Van Aelst and Walgrave, 2004) and the "hyper textual architecture" (Kahn and Kelner, 2003) of the internet and new social movements makes for a symbiotic relationship between new forms of communication.

How Activists Benefit from Changing Media Habits

A Reuters Institute for the Study of Journalism's *Digital News Report* study of 26 countries found that "51% use social media as a source of news each week, with approximately one in ten (12%) respondents using social media as their main news" (Levy, Newman, Fletcher, and Neilsen, 2016). By comparison, their 2015 report reveals 28% of those surveyed use social media as a main source of news. In both studies, Facebook was main social media platform. Technological changes show that now hand held computers (tablet and smartphone) are the main ways of accessing for news sources with "smartphone usage for news is sharply up, reaching half of our global sample (53%), while computer use is falling and tablet growth is flattening out" (Levy *et al.*, 2016). Finally, trust is important with "Trust in news is highest in Finland (65%) and lowest in Greece (20%). Almost everywhere, editors and journalists are trusted less than news organisations" (Levy *et al.*, 2016). Reuters Institute for the Study of Journalism's *Digital News Report* (2015) found that;

> a news journey is less likely to be a brand homepage and increasingly likely to be via a search engine, a social network, email, or the lock screen of a smartphone. We see significantly fewer people accessing the front page of a news website where a list of stories is displayed. More people are going directly to stories via a side door such as search or social media.

Since 2014 news consumption saw an increase of 14%, with the highest growth in the UK, US, and Japan, and the "average weekly usage has grown from 37 to 46% across all our countries. Two-thirds of Smartphone users (66%) are now using the devices for news every week" (Newman, Levy, and Neilsen, 2015). An increase in online video consumption, social media, and new visual formats "coincides in many countries with a fall in audiences for traditional TV bulletins. The trend is most pronounced amongst the under 35s" (Newman, Levy, and Neilsen, 2015), as younger audiences increasingly use Facebook (41%), YouTube (18%) and WhatsApp (9%) for the news. The UK is unique in that over fifty percent (51%) of Smartphone users' source news from the BBC news app, with 25% of those surveyed only use their Smartphone app to access news stories.

Despite social media's perception of being prone to inaccuracies and manipulation, it is a common tool that both aids and compresses protest movements. Social media provides the tools for a voice for activists, through websites, blogs, and tweeting specific journalist and have in some cases learnt about social media from activists. However, as this chapter will show too much emphasis on social media can be to the detriment of any protest. Social media provides a voice, but with a greater chance of inaccurate information, its openness to manipulation, and the vast numbers of groups and people vying to be on the news agenda there runs the risk of losing focus.

Gibson and Gibbs' (2013) study on social media also found that greater knowledge production aided by the transition to Web 2.0, transformed "information dissemination and consumption" (2013, p. 89). Moreover, Gibson and Gibbs (2013) found there were "possibilities of social media 'cutting through' to new and unknown audiences" (2013, p. 88). Bertot, Jaeger, and Hansen (2012) drawing on Pirolli, Preece, and Shneiderman (2010) note that "as the number of users has increased there has been a growing interest in applying social media toward addressing national priorities, not just using them for entertainment or corporate purposes" (2012, p. 30). Adopting of social media sites such as Twitter by government and institutional bodies shows that "as government social media initiatives are launched and evaluated, design lessons can be extracted and shared to achieve these and related goals" (2010, p. 31). The outcome is a growth in e-government practices and an increase in policies that improve the dissemination of information to "communicate with the public, and between agencies" (2010, p. 31).

Although using social media enables collectives to circumvent the traditional gatekeeping practice of press officers and media centres to find their own voice, the hit and run, or tactical, media strategies (Garcia and Lovink, 1997) designed to gain maximum media coverage only work for a brief time. Consequently, any rationale for an action can be lost. In a sea of tweets, requests for likes can be detrimental to public understanding of the aims and objectives of any mobilization. Thus, having set out the role of social media in relation to traditional media practices, this chapter will (1) chart the relations between journalists and activists and (2) explore how social media has changed environmental activism and the media. This chapter seeks to explore what Anderson's identified over twenty years ago that "at present little is known about grassroots campaigners' use of new media technologies such as camcorders, mobile telephones and the internet" (Anderson, 1997, p. 85). In doing so, it will look at citizen journalism, social media tactics, and the use of technology in a cat and mouse game between police, business and the media that's led to some innovative ways to gain media attention. First, this chapter will look at how the internet alters the dissemination of information from a top-down to network power and networked journalism.

Van Der Haak, Parks, and Castells (2012) argues that when information is shared, developed, or added to, the piece of journalism is no longer a single authored piece, tweaked by subs and editors. News reports are now "driven by a networked practice dependent on sources, commentaries, and feedback, some of which are constantly accessible online. The actual product of journalistic practice now usually involves networks of various professionals and citizens collaborating, corroborating, correcting, and ultimately distilling the essence of the story that will be told" (Van Der Haak *et al.*, 2012, p. 292).

Today the boundaries between an individual's experience of environmental injustice, a communities environmental injustice or an entire countries (such as small island nations) are blurred as "individual grievances and feelings are transformed into group-based grievances and feelings within social networks. Networks function as a 'socialization device' as they enable the formation of a mobilization potential and provide or reinforce political awareness around a given protest issue"[3] (Stekelenburg *et al.*, 2015).

Network Power, Network Journalism, and Functions

Network journalism and network power are also useful for protest groups. In a digital age, journalism has also adapted and changed towards a more networked journalism, as opposed to top-down model, but that has not to say any change in the order of discourse. Networked journalism has also affected the various news platforms, and 24 hours news cycle pressures on "professional" journalists. Networked Journalisms is the sharing of information. Rooted in communications theory and analysis, Networked Journalism Networked journalism refers to a diffused capacity to record information, share it, and distribute as

> a world in which information and communication are organized around the Internet, the notion of the isolated journalist working alone, whether toiling at his desk in a newsroom or reporting from a crime scene or a disaster, is obsolete. Every journalist becomes a node in a network that functions to collect, process, and distribute information. (Jarvis, 2006 cited in Van Der Haak *et al.*, 2012, p. 292[4])

Networked journalism is the public participation of information and shaping debates. Similar to a wiki and other new media, networked journalism enables greater horizontality that is a non-hierarchical structure, a rhizomatic way of sharing data, whose only tangible boundary is the digital divide. By sustaining a network model of journalism, there will remain a "continuous relationship between the journalists and the public is what is going to sustain Networked Journalism as a community and therefore, as a business model" (Beckett, 2008, p. 58). Like social movements in order to adapt and survive then networked journalism needs to "mean more than just interactivity it must be considered in the much broader context of changing technology and social behaviour" (Beckett, 2008, p. 47). Thus, there are similarities between the organizational structures of social movements and network journalism, such as a lack of gatekeepers and adoption of new technologies, although journalists remain more powerfull.

Digital media provides a platform where logics thrive, Logics in this example is where "a set of embodied social and cultural dispositions shaped by informational capitalism that oriented actors" (Juris, 2012, p. 9), that protesters challenge through a "collaboration via decentralized co-ordination and directly democratic decision making; and [...] self-directed networking" (Juris, 2012, p. 9). Protesters adoption of social media, websites, and listserv is "characterized by a pervasive cultural logic of networking that arose as activists began to use digital media" (Juris,[5] 2012, p. 9). The autonomous, decentralizing ability of the internet and social media provides a network of communications that first emerges with the Zapatista movement in South America.

Changes in Media Practice: A Return of the Protester's Voice?

Zapatista movement (known as EZLN in Mexico) were one of the first protest groups to use online platforms (pre-social media) to gain public support in their mobilization against the North American Free Trade Agreement (NAFTA[6]). The Zapatista movement found that network power of protest enhanced with online communication strategies, worked well as a "foundation for global networks undermining much of the Mexican state's formal and informal control over production and distribution of information. Slowly, radio, newspapers and television also became less inclined to simply repeat official information" (Krøvel, 2011, p. 131). The Zapatista movement technology and the internet provided a means to disseminating information outside the governmental hierarchy of credibility and gate keeping practices to provide a "broad-based, local and national networks, run by communities, and linked internationally, by the Internet, have proved themselves capable of bringing together very large groups of people in very short spaces of time" (Kingsnorth, 2003, p. 75). A growth in communication technology founded on Internet tools meant collective such as the Zapatistas benefited when "global networks of power and counter-power landed simultaneously to confront each other in the spotlight of the media" (Castells, 2009, p. 340).

Castells (2009), Collier and Collier (2005) and Klein (2002) see the internet as providing a platform for the EZLN to shift power away from journalists and into the hands of activists. Castells' (2009) advocates the Zapatista movement as an example of the internet providing a platform to challenge international policies. Klein argues that "Marcos himself was a one-man web: he is a compulsive communicator, constantly reaching out, drawing connections between different issues and struggles" (2002, p. 217). Klein echoes Castells "the strategic victory of the Zapatistas was to change the terms: to insist that what was going on in Chiapas could not be written off as a narrow ethnic struggle, that is was both specific and universal" (Klein, 2002, p. 217).

Why then do we need traditional journalism if power relations shifts to the point society is based on a network of information that "participants themselves can distribute ... directly to their audiences without the mediation of journalists" (Krøvel, 2011, p. 135). Yet, as Krøvel (2011) notes, it was not the network of information and independent journalism that challenged the government's position, but the physical "caravan" that travelled from Chipas to Mexico City, traversing from state to state, with "journalists from dozens of countries following the caravan" (Krøvel, 2011, p. 136). When the caravan reached Mexico City, some 200,000 people greeted the numerous Nobel Prize winners, authors, artists, and musicians that had travelled with the EZLN. Thus, Castells' interpretation of the ELZN as largely successful due to its virtual presence romanticized the new technological roles within social movements. The ELZN media event occurred due to the physical procession, which garnered media attention over the networks of information. The Internet did play an important supporting role, to disseminate information "so that speeches and announcements could be published almost without delay" (Krøvel, 2011, p. 136).

Berger (2001), Chandler (2004) and Krøvel (2011) argue there were several other factors that bough the ELZN into the world's spotlight. What was problematic with listserv Chipas-L was EZLN found it more useful "when trying to gather information to correct or repudiate claims in the mainstream media" than to disseminate information to supporters and the press (Krøvel, 2011, p. 133). Activists learnt the internet often provided a breeding ground for "spreading unsubstantiated rumours" and that is was "of great importance that the information" put out via the Chiapas-L list serves was "trustworthy" (Krøvel, 2001, p. 133).

Journalists were no longer reliant on the listserv or government rhetoric for information, and could produce their own version of events, meaning, "There was less need for the production of alternative news" (Krøvel, 2011,

p. 137). Journalists from around the globe and a "large number of news outlets [were] also in tow, producing relatively independent and reliable information" (Krøvel, 2011, p. 137). Paradoxically, a U-turn in the application of listservs meant Listserv administrators became reliant on traditional media reports, with every fourth article cut and paste from mainstream media outlets (Krøvel, 2011). Although traditional journalism held more power through their logics (see Juris above), the internet did enable ELZN to challenge the earlier press releases by the state and help develop "robust modes of independent and critical journalism with the traditional institutions of profession news production" (Krøvel, 2011, p. 137). Significantly, despite relying on traditional media for information, the internet aided the flow of information around the globe. The World became aware of the post-modern discourse of the Chiapas, their engagements with NGOs, indigenous peoples, women's rights and was reflective of resistance towards global neoliberal politicalize discourses. Thus, protesters' utilization of the rhizomatic, horizontal architecture of the internet (Pickerill, 2003) through new technologies, smartphones and adaptation of networked journalism, becomes a pivotal tool for organizing protests, informing the media or voicing opinion, and has become a "key ingredient of the environmental movement in the global network society" (Castells, 2009, p. 316).

The World Wide Web provides the tools to enable the activist movements to develop their own media and political strategies, and has extraordinarily "improved the campaigning ability of environmental groups and increased international collaboration" (Castells, 2009, p. 316). Activists are now able to use a new "global communications infrastructure for something completely different, to become more autonomous" (Notes from Nowhere, 2003, p. 65). The capacity of new technologies to support and sustain dispersed coalitions of protestors and new forms of political organization has been witnessed in the anti-capitalism protests (J18 and Seattle in 1999, the May Day protests between 2000 and 2004, the G8 Conference in 2005 and G20 Conference in 2009, and Cop 15 (2009), and CoP21 (2015), social media and the internet equips many individuals and protest collectives and groups with skills to produce citizen journalism.

Citizen Journalism and Capacity Building in the Community

Citizen journalism's first usage or "the first newspaper usage of the term 'citizen journalism' appears in the *New York Times* on 8 September 1976" (Junger,

2014, p. 82) But, it was 15 years later the term came into common usage. Citizen journalism is user generated content (UGC) and that enables activists to invert traditional journalistic practice. Citizen journalism works for social movements and activists by combining professional journalistic practice with the internet and the idea of networked journalism. Citizen journalism also aids many to develop the strategy of outreach, and important way to bypass traditional media.

Protest groups have worked out that by producing citizen journalism it can fill an identifiable gap of that time-poor journalists with 24-hour news demands by providing a ready-made narrative. Australian activists campaigning over the impact of mining on the Great Barrier Reef are aware of the importance of spoon feeding journalist facing continual pressures,

> journalism is an increasingly less profitable industry. There are less journalists and they are often less senior. They are often—this is quite general obviously—but they are more—they have less time. Therefore, if you can do more of the work in putting the story together, they are more likely to take it. That doesn't sort of overcome editorial lines, however there is a lot more you can do if you do the work yourself in packaging it up in an appropriate way and giving them quality stories. They will still run it. (Interview with Get Up! Adrian Dodd, July 2015)

In the UK, VisionOnTv who have been documenting protest movements for many years, now run citizen journalism and video journalism workshops at the Hive, North London. VisionOnTv's Richard Hering says it is more important to build a ground swell of individuals produce footage, instead of sending in "witness contributions" to the BBC. Richard is wary about offering the BBC the same news footage they can provide themselves, he says, "we need to allow ourselves not to be bought-in by the mainstream media. So don't send your clips to the BBC, we should be doing our own distribution, we need those networked in every place" (interview with Richard Hering, September 2011)

> Mass citizen journalism is a way to challenging multinational media corporations such as the BBC. Richard notes, that "in Liverpool we have a group up there, they've made 58, at the last count, films in sort of six weeks, and they're learning, their having meetings, they're teaching each other, it's a self-help group because we can't possibly oversee it from here [London] and so we go on line and give what help is needed". (Interview with Richard Hering, September 2011)

Self-promotion and citizen journalism go hand-in-hand, and climate camp collective's decision to re-engage with traditional journalism was partly to aid citizen journalism. Climate camp's decision to exclude ENGO's meant the camp had less media power to wield and a smaller workforce was a major hindrance for their citizen journalism outputs. In order to have a successful outreach strategy and connect with the local communities around the sites of the camps, without the power and influence of large environmental NGO, there emerged a tension in how best to include all whilst traversing from the outside, in—"And so that was in interesting kind of tension, how do we do outreach? Because you don't want Greenpeace, which is such a powerful NGO coming in, obviously, and dominating proceedings" (interview with Richard Hering, September 2011)

Figure 6.1. VisionOnTv Media Tent—Climate Camp 2007. Image courtesy of the author.

However, citizen journalism can be a challenge to reach a critical mass in the volume of footage and maintaining news interest and agendas' without falling into the social media trap of preaching to the converted. Capacity building and maintaining a large enough volume and on the news agenda can be difficult. Despite the hurdles, Richard feels that if protesters can produce a rapid turnaround news, done by everybody, can go out on the streets tonight and take the latest thing, put a bit of journalism into it in both the description and what you say onto the tape, get that up straight away before the BBC. We need that to be happening everywhere and then we've got a new media and potentially we're contributing to a new society. (Interview, 2012)

Citizen Journalism is seen as vital, because experience with journalists has left some protesters feeling misrepresented. *Guardian newspaper* journalist Paul Lewis's article "Climate Camp gets a lesson in citizen journalism" (28 August 2009, p. 19) sets out the increased use of social media and citizen journalism as a repertoire of protest. The article focused on the use of social media by the activists and the workshops on citizen journalism.

Almost all the quotes in there I didn't actually say. It's very noticeable. I was going nuts (…) his [Lewis's] attitude has changed; he was very dismissive of the radical media (…) He started out his career as a politician, dissing radical media to try and get himself onto this greasy pole, which he did make it up the top of, impressive. Doing a good job on the top of the greasy pole but actually completely misquoted just about everything.

Protesters and individuals could consider becoming their own investigative journalists. With more time-friendly access, the opportunity to exploit the gap in time-poor mainstream journalism and the network power that social media provides, means as one Rising Tide stalwart says,

I think that ourselves, we need to do our own investigative journalism (…) I mean, for the Drax camp, we had a long, really long, difficult debate in Rising Tide as to whether we should fly over two activists from the US, who were going to set up Rising Tide and the Climate Camp movement in the US. Eventually we did and, obviously sometimes you've gotta take a chance. Anyway, they went back and within six months, they had set up a 200-group network, in six months. And the way they did it was by tapping into existing networks, mainly of native communities, mainly of native communities that were already engaged in fighting something that they didn't see as climate change, like mountain top removal for coal but was climate change and was literally as grass—roots as you can get; and not just join these up within a Rising Tide, climate change, direct action network but put them in touch with each other. Amazingly, they weren't even in touch with each other (interview with Rising Tide activists, J, September 2011).

Citizen journalism combined with outreach to communities near a site of contestation can also be a powerful strategy for activists. Outreach is a means

Figure 6.2. Climate Camp Citizen journalism guide. Image courtesy of Hamish Campbell and Richard Hering at VisionOnTv.

to getting the word out about a campaign and connecting with people who would be interested. It can also be used to increase pressures on politicians, communities and spread information. The best public outreach gets people to understand and care about your campaign.[7] As Richard says,

> on the other side we are trying to do outreach. One of the most interesting studio shows that we [VisionOnTv] did was between two activists, George Marshall of Rising Tide and Uri Gordon who had written a book on Anarchist political theory and one was saying, "We need Trade Unions here, working class people, not just because they happened to live locally and they're interested in the campaign, but because they come from Manchester, or wherever" RH. (interview, 2011)

Citizen journalism, investigative journalism and outreach all provide grassroots networks with a voice, but technology enables others ways to challenge the eco-governmentality and criminalization of ecoActivism (see Chapters Three and Four). Despite the digital divide, and large percentage of middle-class coordinating protests, there is a realization that there is no alternative now. The alternative media and communication channel of listserv and Indymedia no longer serves its purpose. With easy access to journalists and the public via Twitter, Facebook, Pinterest, Instagram, and Flickr, there is a sense that despite social media faults, if offers more, so long as those using it are aware of the limitations. VisionOnTv's HC says, "The success of the corporate stuff comes at the price of the failure of the radical alternative stuff and the radical, alternative media is in the doldrums nowadays. I mean, Indy Media almost doesn't exist" (interview with Hamish, September 2011).

Activists come to realize that social media is useful when aware of its limitations and susceptibility to opening up any action to police surveillance. Activists accept that challenging what has become, "The best tools available, the most attractive or the tools everybody knows, not necessarily the best but the ones everybody knows and the one all your friends are on it's an impossible argument to win" (interview with Richard, September 2011), and can be heavily influence by the instantaneous nature of social media, often outweighs the risk of criminal prosecution.

During a mobilization at Aberdeen airport Twitter and other social media sites brought greater benefits to circumvent traditional media, especially when the police attempt to block out action. Seven Plane Stupid's took direct action on the apron of Aberdeen airport, Scotland (3rd March 2009).[8] The activists built a cage and lock themselves in, in protest against carbon emissions from short haul flights. The cage was surrounded by two fire engines, in what activists believe was an attempt to stop traditional media coverage "You know, one of the things about the Aberdeen protest, they put a fire engine in front of the cage that we were in so that no media could see us" (interview with Dan, June 2014)

Social media tactics offer speed, getting images of a mobilization into the public domain before any authority can react to an action. Images can be uploaded and onto the internet, so when preventative measures are taken, "by that time, we'd taken our own photos and sent it out to [say to the police] 'I've done it anyway,' so it's been really useful" (interview with Dan, June 2014). Using "Twitter on actions that we've done, just being able to upload something we've occupied, Manchester, Stanstead, Aberdeen airports, it's been genius. It's been absolutely brilliant in many ways for organizing actions and bypassing, A, the police and B, the powers that be, and C, the traditional media" (interview with Dan, June 2014).

With the digital divide, limited capacity for network building across media zones, small network support—preaching to the choir—and datamining smartphone to track and trace individuals shows that a heavy social and online media presence can be detrimental to long term movement success. Social media can be problematic for protesters, for four main reasons. Firstly, it is monitored by the state, so any plans for protesters made on social media are often used in court cases as evidence, and removes the element of surprise—a strong necessity for successful direct action. Secondly, it is prone to misinterpretation and illusionary tactics. With the exception of Tahir Square in Egypt and the Arab Spring Uprising (2011) (cf. Gerbaudo, 2012), a thousand likes and hits does not often translate into thousands of people on the streets. Thirdly, social media is the predominant preserve of the middle classes and the digital. Fourthly, too much emphasis on organizing through social media can often mean a loss of focus for the event. When focus is lost, so is the purpose of an action, and protests are open to ridicule by the media and public.

A continual use of social media and websites to communicate strategies and plan demonstrations leaves any individual or collective open to criminal charges. Social media might be, as one activist says the "best organizing tools we've ever had but they are also prisons and we end up in prison because of them". (interview with John, 2015) Indeed, tweets were part of the evidence brought against G20 activists. During the trial of activists arrested at the G20 meeting in Pittsburgh, USA (2009) activists communicating via social media and mobile phones were traced and arrested. One activist there retells the tale, "at the G20 in Pittsburgh that the government just traced the sender and was able to arrest them in some hotel" (interview with MC, July 2015). The same year, an activist arrested at the United Nations Climate Change Conference of the Parties (CoP 15), Copenhagen, Denmark was presented at trial with evidence of participation from hundreds of Tweets. An activist at Copenhagen (2009) notes, "Then you've got the CoP 15, where the people were arrested and they came to trial, they had a stack, like six inches of every tweet, and the CIA has actually set up because 'Green is the new Red,' set up a sub-office of the CIA, to follow the climate change [campaigners]" (interview with MC, 2011). Social media remains the established mode of communication, despite it also being a "universal tool of surveillance of the people we're fighting against" (interview with Hamish, 2011).

Hamish and other stalwart activists see that continual reliance on website present for online branding and social media means many activists are openly providing police with the information they need, so that "activists relying on them wholesale with such national naivety. I think there's an act of criminal naivety happening in this digital space where everything's open, that everyone is jumping into prison" (interview with Hamish, 2011). At the same time, there is a realization of the organizing potential of social media

> Don't get offline yourself because you lose a huge amount of organising power, build tools which their *purpose* it to get people offline, we need to concentrate on that. We've got to get people out of the hamster cage, so we've got to build better tools, tools which work in ways which people would like to work. On-line openness is the only solution, so stop building secure online communities because they're fantasies, there is no security on-line, it's digital, all information just wants to be free. (Interview with Hamish, 2011)

Some attempt to challenge or bypass by removing phone batteries blocks surveillance signals is outmoded as technology has developed to counter the problem. Using unregistered laptops and encryption email software has some effect, but in general, activists that continue to use social media and the internet can be traced to the frustration of some activists. Individuals realize there is potential in social media, as one asks? "How can we use social media to try and overcome those barriers?" (Interview with Richard, September 2011). There is an awareness that not all protesters engage with social media, as Richard adds

> Just as indigenous communities, and low socio-economic communities are less engaged with social media, and media strategies is because they are living with the

> consequences of climate change on a daily basis an too busy adapting and dealing to be making videos, tweeting their views, or organizing swarms or week long protest camps. (Interview, with Richard, 2011)

Once described as the middle-class suburban rebel observers of 1960s and 1970s counter-culture groups, Hall (1968), and Heath and Potter (2004), observes how the drop out, tune-in of 1960s Timothy Leary is only possible for a middle-class looking for an escape from middle-class suburbia. A middle class component of many of the movements gives a much stronger focus on social media, new science, and the latest technologies. Class and social mediacentric reporting has contributed to division within movement between stalwarts and novices, no more so with the emergence of social media and a reinvigoration of protests—as one stalwart says "2005 became the cut of point when it became less altruistic and the middle classes joined, and then you have all of the self-promotion, and using all the media out there" (interview with Michael, 2011).

An activist from So We Stand, an environmental justice collective working with communities directly affected by climate injustice such as women and low socioeconomic groups sees social media and the middle classes as hand-in-hand, and "a broad statement about new media would be that I think that it's very useful in middle-class, white activism and preaching to the converted" (interview with activists, NR, September 2012). Whereas in contrast the traditional, mainstream media is offers greater inclusivity with a wider representation for grassroots and marginalized groups—they say "traditional media has the capacity to break down those barriers more than social media because I think the way that social media is used is different for different demographics, whereas everybody uses traditional media for news feeds, social news and political" (Interview with activists, NR, September 2012).

Consequentially too much emphasis on social media, image events and mediacentric practices to mirror network journalism, is a loss of focus in why a mobilization, protest or single issues event is necessary. Once a protest loses focus, then the public have less understanding of what the action is attempting to address. As JS says, "I think when things begin to lose their focus and they turn in on themselves and that also has consequences for how it is covered by the media" (interview with John Stewart, 2011).

Earlier collectives of the time were prone to loss of focus leading to many collective disintegrating and others finding new paths or causes. For example,

Reclaim the Streets (1990s) dismantled, when many wanted to focus more on the environment went to form London Rising Tide environmental groups, whilst other engaged in their passion for art. As stalwart MC says, "In the long-term, Reclaim The Streets was diverse in a sense but then again, when other people tried to introduce things, later on, like Indy Media or Wombles, and that sort of thing, when it wasn't agreed upon, then the people who proposed all this went off and set up their own groups" (interview with MC, September 2011).

Whereas in the Climate camp movements, because of the number of different collaborators and groups involved it may be harder to start up elsewhere, and there is a difference between having a focus and being single issue, you know. And if we are talking about climate change, it is a kind of issue about scale, so it is very interesting where we need to say, we need a focus, a clear focus, while keeping the hugeness of it (interview with Hamish, September 2011). Stalwart activists Steve, backs up Hamish's claims, he says that:

> Most activists agreed that the climate camps began with a clear focus, "climate Camp a bit. It had a clear focus to start with, it had a focus on Drax, it had a big focus at Heathrow on aviation, it had a focus at Kingsnorth on coal, and another overall focus was on climate change. But by the time it got to Blackheath it was much less clear what the focus was". (Interview with Steve, September 2011)

In retaining a focus, then "people's divisions don't matter. So you can define yourself as militant or not, you know, you can define yourself as moderate, but you're all working towards the same end, you are all working towards the same goals, not the same thing. The difference between having a focus and being single issue, but, and everybody can sort of safely agree on that" (interview with Richard, September 2011).

Lovell and Doherty, Plows, and Wall (2003) generally argue that, for activism to be successful, the focus should not be on influencing the news media discourse; but the measure of success or failure comes by how influential the event will be on changing political and social ideology. Direct action is not about media coverage—it is about the ability to take back power, because the "mediatisation of a protest should not be used as a mark of success or failure, but about taking a bit of control back into my life that's been taken away from me by the State, corporations or whatever" (Doherty *et al.*, 2003, p. 670) It is worth remembering, "media mediation is [not] necessary for a protest to be worthwhile" (Doherty *et al.*, 2003, p. 674); instead, targeting companies and organizations, using direct action and civil disobedience as a preventive measure is, for Doherty *et al.* a more effective means of changing social and political ideology. Direct action and physical protest are more effective as targeting a company through direct action can "have a direct effect on a company's profits (…) and an indirect one if consumer and investor confidence in a company means a drop in share price" (Doherty *et al.*, 2003, p. 676), with the role of the media being negligible.

Going Online to Get Offline

Environmental activists can "increase their chances of enacting social and political change—even if they start from a subordinate position in institutional power, financial resources, or symbolic legitimacy" (Castells, 2009, p. 302). As HC notes, social media "are the best tool we have ever had", but he is very cautious about embracing social media:

> I think, activists relying on them wholesale with such national naivety that I want to scream and jump … and the success of the corporate stuff comes at the price of the failure of the radical alternative stuff and the radical, alternative media is in the doldrums nowadays. I mean, Indy media almost doesn't exist. (Interview with Hamish, 2011)

According to Morozov "the problem with political activism facilitated by social networking sites is that much of it happens for reasons that have nothing to do with one's commitment to ideas and politics in general, but rather to impress one's friends" (2011, p. 186) The internet shouldn't be a solitary force for mobilization of new political forms, but one factor of it, and "success is conditioned by many factors that have little to do with the internet" (Gavin, 2009, p. 130). Gavin (2009) observes that if traditional media sees the web as an add-on "an alternative transmission mechanism for these media, which are far and away the public's most trusted and most often used source of political information" (Gavin, 2009, p. 138).There's a strong realization that as S adds that there is a need to "Do more face-to-face stuff and, whatever means necessary, we are going to go to the actual mainstream media is, with all the growing things that are coming, we need to continue to expose the role of the state" (interview with Steve, September 2011).

Activists need to be more aware and careful in light of state surveillance, and be aware of how information is shared, especially in light of the PC Mark

Kennedy incident. As HC says, "please activists think about security models and what's appropriate ... we've got to build better tools so that activists, so that real people might actually want to use our media and might want to, use these tools we build." (Interview with HC, September 2011).

Furthermore, continuing to organize and coordinate via social media opens up groups' to surveillance and manipulation of events and journalists are able to attend either by identifying themselves or not. At each meeting, it has made clear whether is it open or a request for mobile phones to have the batteries removed to prevent the police and others from listening in on meetings. As an activists says the "state are complicit in this scheme with the media and with multinational corporations, and the power that is out there isn't held by governments, governments just set the parameters and if multinationals don't like it in that little parameter, they'll move somewhere else into another parameter" (interview with Chris, 2011).

The internet and social media networks are a good reorganizing tool, although "what people have lost in the social media hype, is the ability to have a chat. Too busy, got 3000 email to answer. No, no I'm not friends with you because you're not on Facebook" (interview with Dan Glass, September, 2011). As Morozov observes:

> While Facebook-based mobilization will occasionally lead to genuine social and political change, this is mostly accidental, a statistical certainty, rather than a genuine achievement. With millions of groups, at least one or two of them are poised to take off. But since it's impossible to predict which causes will work and which ones won't, Western policymakers and donors who seek to support or even prioritize Facebook-based activism are placing a wild bet. (Morozov, 2011, p. 180)

Conclusion

Changes in our consumption and production of news stories are impacted by new media technologies. Smartphone, tablet and a handful of apps makes it simpler to consume news. Editing apps and packages on portable devices aids citizen journalism. With citizen journalism comes the opening up opportunities for independent media to produce their own news stories to upload to TV and radio companies. New media technologies have altered journalism into a network power providing more access from the local to the global as the Zapatista case study shows. At the same time, the internet and social media fails to produce a promised equilibrium between grassroots movements and the top-down news production of the traditional media. Too much focus on social media strategies, hits and likes, can lead to a loss of focus on the objective and too much focus on a web presence. The next chapter looks at other ways to keep a balance between retaining the object for any action/activism whilst having a media presence.

Notes

1. LGBTQI refers to the Lesbian, Gay, Bisexual, Transgender, Queer and Intersex persons, and people who do not adhere to sexual and/or gender norms use the term to self-identify in a positive way.
2. See http://blogs.lse.ac.uk/politicsandpolicy/how-effective-are-protests/?utm_content=buffer-9d295&utm_medium=social&utm_source=twitter.com&utm_campaign=buffer
3. See http://blogs.lse.ac.uk/politicsandpolicy/how-effective-are-protests/?utm_content=buffer-9d295&utm_medium=social&utm_source=twitter.com&utm_campaign=buffer
4. See http://ijoc.org/index.php/ijoc/article/viewFile/1750/832
5. See https://www.researchgate.net/profile/Jeffrey_Juris2/publication/263133968_Reflections_on_Occupy_Everywhere_Social_media_public_space_and_emerging_logics_of_aggregation/links/53e7f8000cf21cc29fdc2e1e.pdf
6. Canadian, Mexican, and American governments drew up the North American Free Trade Agreement (NAFTA), framing it as an opportunity to lower trade barriers and provide opportunities for export. In reality, it led to lower subsidies to the indigenous populations, whilst opening up opportunities for large corporations. To appease the indigenous population, the Mexican government agreed to an amendment to the Treaty, and when the newly elected president, President Fox, sent the Indigenous Rights Bill to be passed in 2001, the Zapatista army (a group of farmers) travelled the 2,000 miles to the capital to address Congress. When they reached Mexico City, 100,000 people greeted them. The mobilisation of thousands of people through the internet showed the organisational potential of the World Wide Web. By 2001, details of the Zapatistas' protest against NAFTA had spread around the world via email, websites, and blogs.
7. http://rabble.ca/toolkit/guide/public-outreach-activists
8. See https://www.youtube.com/watch?v=WAtfzITYGh0
9. http://environmental-justice.com/so-we-stand/

References

Anderson, A. (1997). *Media, Culture and the Environment*, London: Routledge.

Beckett, C. (2008). *SuperMedia: Saving journalism so it can save the world / Charlie Beckett*. Malden, MA: Wiley.

Beckett, C., & Mansell, R. (2008). Crossing boundaries: New media and networked journalism. *Communication, Culture and Critique, 1*(1), 92–104.

Berger, M. T. (2001). Romancing the Zapatistas: International Intellectuals and the Chiapas Rebellion. *Latin American Perspectives.*, *28*(2), 149–170.

Bertot, C. J., Jaeger, P. T., and Hansen, D. (2012). "The Impact of Polices on Government Social Media Usage: Issues, Challenges, and Recommendations", *Government Information Quarterly*, 29, pp. 30–40.

Bowles, N., and Nadon, D. (2013). *Staging Social Justice: Collaborating to Create Activist Theatre*, Carbondale, IL: Southern Illinois University Press.

Castells, M. (2009). *Communication Power*, Oxford: Oxford University Press.

Chandler, D. (2004). "Building Global Civil Society 'From Below'?", *Millennium*, 33(2), pp. 313–339.

Collier, G. and Collier, J. F. (2005). The Zapatista Rebellion in the Context of Globalization. *The Journal of Peasant Studies.*, *32*(3/4), 450–460.

Couldry, N. (2003). *Contesting Media Power: Alternative Media in a Networked World*, Lanham, MD: Rowman and Littlefield.

Doherty, B., Plows, A., and Wall, D. (2003). "The Preferred Way of Doing Things: The British Direct Action Movement", *Parliamentary Affairs*, 56, pp. 669–686.

Garcia, D., & Lovink, G. (1997). The ABC of tactical media. *first distributed via the nettime listserv.*

Gavin, N. (2009). "The Web and Climate Change Politics", in T. Boyce & J. Lewis (eds.), *Climate Change and the Media*, London: Peter Lang, pp. 129–145.

Gerbaudo, P. (2012). *Tweets and the Streets: Social Media and Contemporary Activism*, London: Zed Books.

Gibson, C., and Gibbs, L. (2013). "Social Media Experiments: Scholarly Practice and Collegiality", *Dialogues in Human Geography*, 3(1), pp. 87–91.

Hall, S. (1968). *The Hippies: An American 'Moment'*, Stencilled occasional paper, Centre for Contemporary Cultural Studies. SP No. 16. Birmingham: University of Birmingham, Centre for Contemporary Cultural Studies.

Heath, J., and Potter, A. (2004). *Nation of Rebels: Why Counterculture Became Consumer Culture*, New York, NY: HarperCollins.

Hutchins, B., & Lester, L. (2006). Environmental protest and tap-dancing with the media in the information age. *Media, Culture & Society*, 28(3), 433–451.

Hutchins, B., and Lester, L. (2009). "Power Games: Environmental Protest, News Media and the Internet", *Media, Culture and Society*, 31, pp. 579–595.

Juris, J. S. (2012). "Reflections on #Occupy Everywhere: Social Media, Public Space, and Emerging Logics of Aggregation", *American Ethnologist*, 39(2), pp. 259–279.

Jurrat, N. (2011). Citizen Journalism and the Internet. *Open Society Foundation: Washington.*

Kahn, R., and Kelner, D. (2003). "Internet Subcultures and Oppositional Politics", in D. Muggleton and R. Weinzierl (eds.), *The Post-subcultures Reader*, London: Berg, pp. 294–309.

Kawamoto, K. (Ed.). (2003). *Digital journalism: Emerging media and the changing horizons of journalism*. Rowman & Littlefield Publishers.

Klein, N. (2002). *Fences and Windows: Dispatches from the Front Line of the Globalisation Debate*, London: Flamingo for Harper Collins.

Kingsnorth, P. (2003). *One No, Many Yeses: A Journey to the Heart of the Global Resistance Movement*, New York, NY: Free Press.

Krøvel, R. (2011). "The War in Chiapas: The Fall and Rise of Independent Journalism", in A. Charles and G. Stewart (eds.), *The End of Journalism: News in the Twenty-First Century*, Oxford: Peter Lang, pp. 129–138.

Laser, R., Erickson, L., and Kessler, J. (2010). "Moving the Middle on Marriage: Lessons from Maine and Washington", *Third Wat Fresh Thinking*. Available at http://content.thirdway.org/publications/248/Third_Way_Report_-_Moving_the_Middle_on_Marriage.pdf

Lewis, P. (2009). "Climate Campers Get a Lesson in Citizen Journalism", *The Guardian* (London), August 27, p. 19.

Levy, A. L. D., Newman, N., Fletcher, R., and Neilsen, R. (2016). Reuters Institute Digital News Report, Reuters Institute for the Study of Journalism. Available at http://www.digitalnewsreport.org/survey/2016/

Morozov, E. (2011). *The Net Delusion: How Not to Liberate the World*, London: Penguin.

Newman, N., Levy, D., and Neilsen, R. (2015). Reuters Institute Digital News Report 2015 Tracking the Future of News, Reuters Institute for the Study of Journalism, UK: Kindle Edition. Available at http://www.digitalnewsreport.org/survey/2015

Notes from Nowhere (Organization). (2003). *We are Everywhere: The Irresistible Rise of Global Anticapitalism*, New York, NY: Verso.

Pickerill, J. (2003). *Cyberprotest: Environmental Activism on-Line*, Manchester: Manchester University Press.

Pirolli, P., Preece, J., & Shneiderman, B. (2010). Technology-Mediated Social Participation (Cover Feature with 7 articles). *IEEE Computer*, 43(11), 20–67.

Van Aelst, P., and Walgrave, S. (2004). "New Media, New Movements? The Role of the Internet in Shaping the 'Anti-globalisation' Movement", in Donk, W. (2004). *Cyberprotest: New media, citizens and social movements / edited by Wim van de Donk, Brian D. Loader, Paul G. Nixon, Dieter Rucht; foreword by Peter Dahlgren*. London: Routledge.

Van Der Haak, B., Parks, M., and Castells, M. (2012). "The Future of Journalism: Networked Journalism (Special feature) (Essay)", *International Journal of Communication (Online)*, 2923(16).

Van Stekelenburg, J., Roggeband, C., and Klandermans, B. (2015). The Future of Social Movement Research: Dynamics, Mechanisms, and Processes. *Contemporary Sociology*, 44, pp. 724–725.

· 7 ·

HETEROTOPIAS

Retaining Power in the Space of Protest

> The use of physical space at climate camp is interesting between the radical media and the traditional media team. At the first one, they pretty much shared the same space, it was pretty low key; … and the fourth one they just didn't see each other at all.
>
> —interview with activists, HC (2011)

The Arab Spring (2011) was a series of protest across North Africa and the Middle East as a rejection of political systems and authoritarianism. The Occupy Protest in countries around the world followed the Arab Spring actions. Protesters from across the globe came together in public squares and spaces. People reclaiming public spaces as an outward demonstration of disunity at the political situations, were affectionately named in the *New York Times* newspaper as "square people" (Friedman, 2014). Square people were more than just a gathering of like-minded individuals but "people who occupied the squares got a sense of power that was absent in the voting booth" (Krastev, 2014, p. 16). Power came from a sense of distrust and disillusionment of the state leading to many of the protest of 2011 and beyond. We can understand the demonstrations as part of the "ongoing principles of socialist anarchist thinking" (Downing, 2003, p. 248). The protest focuses on being "movements over institutions—a distrust of the state", and does so by creating temporary autonomous zones' and places "located to direct action" (Downing, 2003, p. 248). By creating a space from which is be both autonomous yet engaging in the political processes can be understood as a Temporary Autonomous Zone (Bey, 1991) or a Heterotopic space (Foucault, 1986). This chapter discuss how ecoActivists and other protest groups can utilize the space of protest as an outward projection of power to the public and politicians. Where there is strength in a movement is the physical space it occupies. The space of the protest holds power, so long as there is a rationale then power can sit within the space. A protest camp, March, or sit-in generates a space where boundaries of representation (see Ch. 2) shift as power relations between the outside and inside develop. The protest space becomes a political commentary and a mirror for alternative living. As this chapter will show, the outside becomes the inside, the space is not somewhere activists go to challenge the discourse, it is where ecoActivsts power sits for the movement.

We can better understand the role of power in protest space through Foucault's concept of Heteropia (1986). A heterotopias is a "real place, places that do exist, and are formed in the very founding of society—which are something like counter-sites" (Foucault, 1986, p. 24). Counter-sites sit "outside of all places" (Foucault, 1986, p. 24) and exist in a space between a utopia and the real. The space acts as a mirror to the real world, made from ideas in the real. Yet, the mirror transposes neither a true nor utopian reality, but an alternative. For example, a circus hall of mirrors, the image is known, a person, an object, and therefore we can cognitively understand, but the image is also different, alternative, distorted to give another way of seeing. In looking at the climate debates through the heterotopic space of protest, is similar to seeing the space as a mirror held up to show how existing society works, but as a slightly distorted, opaque image, of an alternative. However, a heterotopias is neither a fantasy place, nor completely real—it is "connected with all the space that surrounds it, and absolutely unreal" (Foucault, 1986, p. 25). However, when the space no longer connects with its surroundings, both internally or externally it ceases to work as an alternative.

Feigenbaum, Frenzel, & McCurdy (2013) are not convinced that Foucault's short discussion on heterotopias can be a claim put to protest. Instead, a protest space, is more the position of "utopian politics aimed at radical alternatives to the status quo often inspires protest campers, since their practices are—without doubt—located in the here and now" (2013, p. 220). Feigenbaum *et al.* (2013) make a valid point and some protest, particularly single issues that tend to focus more on a utopian ideal than a heterotopia. Whereas others saw the anti-road movement as "a heterotopic space in which it becomes possible to express a new way of life" (Doherty, 1999, p. 13). A heterotopia is measured and semi-organized whereas a "temporary autonomous zone" (TAZ[1]), are spaces that exist as individual islands in society, a "mini-societies living consciously outside the law and determined to keep it up" (Bey, 1991, p. 1). Hence, as this chapter will show, by identifying between a TAS and heterotopia, its' clearer to see power relations between the protest space and the politic.

Square People and Temporary Autonomous Zones

Public squares working as sites of contestation created communities with "participants often making their protest activities a basis of their political identity" (Krastev, 2014, p. 16). Equally, there was a fear, generated from those outside the protest, so that "protests succeeded in fuelling the anti-institutional imagination of some, they fuelled other citizens' fears of chaos and anarchy, allowing governments to portray the protesters' urge for direct democracy as a threat to public order" (Krastev, 2014, p. 4). Furthermore, the square protests were "unlike elections, protests were able to represent effectively the intensity of public sentiment, and in country after country hostility to elites was at that sentiment's heart. The protests showed that things could change" (Krastev, 2014, p. 12).

Activists might argue the space of protest is more a "temporary autonomous zone" (TAZ) (Bey, 1991), than a heterotopia. There are many similarities between the two, with both a TAZ and heterotopias acting as alternative space. Both explore liminal space that exists in the sensory threshold between two states of being. A TAZ is "like an uprising which does not engage directly with the State is a guerilla [sic] operation which liberates an area (of land, of time, of imagination) and then dissolves itself to re-form elsewhere/elsewhen, before the State can crush it" (1991, p. 3). TAZ are physical spaces that "exist within a world of pure space, the world of the senses" (Bey, 1991, p. 6). Any TAZ functions through a combination of information and desire in order to "fulfil its adventure (its "happening"), in order to fill itself to the borders of its destiny, to saturate itself with its own becoming" (Bey, 1991, p. 6). To survive, the TAZ needs to be ephemeral, whereas a heterotopia space centres on the capacity to build power.

Heterotopic Space and Political Power

There are five main principles of heterotopias: (1) crisis, (2) determined function within society, (3) several sites encapsulated into one space, (4) a slice in time (heterochrony's), and (5) openness to manipulation and alternative directions (Foucault, 1986, pp. 24–26). The first principle of heterotopias is crisis—a space occupied by people who live in society but who are in a constant state of crisis. For Foucault, those in crises are "adolescents, menstruating women, pregnant women, the elderly" (1986, pp. 24). A crisis heterotopias is a space that is outside of society, but at the same time held up as a mirror to show the flaws and discrepancies in any system. For example, the Calais Jungle in Northern France was a crisis, but also holds up a mirror to the socio-cultural and political decision-making process around migration rules and asylum seekers. The camp becomes a space outside of society, but still within a societal system (the EU, France, and UK).

The heterotopias (the camp) also plays a role of having "determined function within society" (the second principle of heterotopias), and that "precise and determined function within a society, and the same heterotopias can, according to the synchrony of the culture in which it occurs, have one function or another" (Foucault, 1986, p. 25).

A determined function of a protest camp is one to both highlight an issue, and act as the conduit. Depending on the cultural or political context, the framing of each movement is dependent on different functions over time. As we saw in Chapters Two and Five, the media focus on Daniel Hopper (a.k.a Swampy) was a moment that led to lessons learnt in future mobilizations, and a policy of rotation, outreach through community engagement; the Critical Mass court case was its determinant function to challenge the tactic of law (see Chapter Four); or the climate camp movement's determinant function to spotlight police surveillance and intrinsic links between neoliberals and the climate debate. Each occasion works as a particular function, to raise wider questions, to examine reasoning, techniques, and decision-making processes that shape public understanding. The physical space itself is part of the process, as it holds power both inside and outside of the action. Unlike marches, and rallies, a lot ecoActivism occurs at the site of contestation. Each cycle or heterochron at the site of contestation. The constantly evolving and increasingly politicking of longer standing movements, those that have built up a culture over 40 or more years, and often combine stalwarts and novices, which is why protest fits well into the heterotopia principles.

An inside/outside dichotomy exist and is important for protest survival. Protests are always open to manipulation and internal fractions that exposes the fifth principle "openness". Once exposed the inside/outside dichotomy, power, and objectives are lost. Rosenwein and Little (1998) warn that inside/outside creates friction between those based long term at the camps (inside) and those who visit (outside). As we will see manipulation and loss of focus from the fifth principle openness creates problems can be as damaging as the role of the state and police, causes antagonisms within a protest movement.

Space and Liminoid Practices of Protest

Within the space of ecoActivism (such as camps and social spaces/squats), cultural practices are built up from repertoires of protest to create maps of meanings within a movement. For many activists, the identifying of a site of protest is part of an ongoing acknowledgement of the historical importance of trying to create a heterotypic space. The space of a protest and where and when a mobilization takes places is as significantly important as too why. Feigenbaum *et al.* (2013) say that;

> Protest camps are not simply mirror images of a social order that is already in place, but nor are they faraway utopias devoid of any relationship to the here and now. Indeed, in protest camps the tensions between antagonistic demands for a new beginning meet with the practical requirements of making things work in the present. (2013, p. 221)

By understanding protest camps as a heterotopic space (Foucault, 1986; Hetherington, 1996); it contextualizes them as alternative spaces. Heterotopias acts as an alternative, although not utopian, space. Hetherington draws on the new age traveller movement to show how "the nomadic New Age Traveller who travels to sites such as Stonehenge [is] engaging in a liminoid rather than liminal ritual process of identity transformation" (Hetherington, 1996, p. 37). Moreover, "the use of heterotopic spaces, such as Stonehenge, which act as sites of social centrality for those who attempt to transform their identities through transgressive and carnivalesque performance … and through the shared sense of belonging that underlies their identification with one another" (Hetherington, 1996, p. 39). The shared sense of belonging is achieved through liminoid practice to help define the identity of protests as a heterotopia.

> The festival is a liminoid social occasion in which the norms and values of society are overturned and a new code of behaviour established such festivals have generally been held on common land sites, the sites of old medieval fairs or at a pagan site like Stonehenge. It is the sites of such festivals that can be described as heterotopia. (Hetherington, 1997, p. 42)

Environmental protests and movements are made-up from a plethora of politically diverse collectives. The multiplicity of movements exists within one space as heterotopias are "capable of juxtaposing in a single real place several spaces, several sites that are in themselves incompatible" Foucault's third principle of heterotopias (1986, p. 26). An example would be the outreach at the Heathrow Climate camp and Plane Stupid's Adopt-a-Resident scheme. The third principle shows why some protest and protest camps also work as heterotopias for the multiplicity of movements, which exists within one space, With a heterotopias being "capable of juxtaposing in a single real place several spaces, several sites that are in themselves incompatible" (third principle of heterotopias) (Foucault, 1986, p. 26).

Gathering many groups under one objective, one focus, allows the heterotopias to work at full strength (heterochrony). Everyone then will "arrive at a sort of absolute break with their traditional time" (Foucault, 1986, p. 26). The protest becomes a unifying space that, regardless of external factors, is united though a common objective, such as economic solutions to the climate debate, a coal seam gas site, an airport expansion, a new supermarket, anywhere that ecological damage is being done, or decisions are made that may lead to ecological damage and climate injustice.

Community Outreach and Protection of the Protest Space

Mirroring the third principle or multiplicity, Plane Stupid shared consensus practices with local residents near the Kingsnorth camp. HACAN Chairman, John Stewart (2012) notes. "they adopted our working practices like consensus decision making" and adding that "it was amazing, we've never had meetings like this, they are so fast, so efficient, you know, they really loved that actually, our working practices" (interview with John, 2012). Through outreach with the local community the Climate camp team were able to both protects their space and that of the protest camp. In the case of Heathrow climate camp, it was the sharing of liminoid practices through an Adopt-a-Resident project.

Adopt a Resident was a strategy to prevent any divisions between the protest camps and local communities. Negative media coverage was often a forerunner for creating tensions between communities and the camps. Activist groups realized there need to be a coming together to dispel any media myths. One scheme was the adopt-a-resident project. Activist "adopts" a resident to teach them the ways of activism. The "Adopt-a-Resident" strategy brought together 40 local residents from Harmondsworth village, and 30 activists. Supported by local MP John McDonnell, activists were paired with residents as a "way of getting round people who were more radical being marginalised (…) so the residents asked to be adopted by activists, and we would do things together, like Ceilidhs and tea parties" (interview with Dan, September 2011). Adapting the repertoire of tactics such as locking-on, they held a "locking on limbo" night—when residents and activists "limbo"—"you had these 60–70 year olds doing a limbo underneath an arm-tube lock-on … to reframe the debate," (interview with Dan, September 2011). Lock-on limbo was a tactic to defuse any divide and rule reporting, and challenge the stereotypical images of environmental activists, by replacing it with images of limbo dancing grandmothers. HACAN spokesperson John Stewart notes, "Linda, a 60 year old, grand-mother, she's certainly not a militant, you know, so that … dispels it" (interview with John Stewart, September 2011).

Connecting with local residents also helps demystify the idea that the space will be one of public disorder. Plane Stupid's John Stewart says "the long term impact on the local campaigners, not just at Sipson, but in West London, certainly once we got involved, people were much less scared of being identified as radical campaigners in the past" (interview with John Stewart, September 2011). In the Heathrow campaign some years before the Climate Camp, we set out to have as broad a coalition as possible and this way of getting round people who were more radical being marginalized. John explains that

> The residents asked to be adopted by activists, we would do things together, like Ceilidhs and tea parties. Adopt-a-Resident was a simple project where activists adopt a local resident, and share knowledge, tips on non-violent direct mobilization, and civil disobedience, media movements, media tactics, and social media engagement. (Interview with John, September 2013)

Activists shared knowledge of old school protest tactics of locking on or die-in. The objective is to alter public perception of a protester, an activists need not be the stereotypical Swampy characterization favoured by the news media.

Adopt-a- resident was one of many links between "different networks and campaigns [that] share the same struggles" consolidated "the ground from which anti-globalization mobilization springs ... through continued focus of generating and building capacity at local group level; eco camps are useful way of doing this" (Plows, 2006, p. 26). Plows (2006) study found evidence of capacity building at a road protest at Blackwood, South Wales (February–April 2004). The action was over roughly two miles of road, costing £56 million pounds, and scheduled to destroy an ancient, and extremely beautiful, stretch of woodland. Over the 11 years, local campaigners had contested the scheme through the usual legal channels, and 11,000 people signed a petition (Plows, 2006). Furthermore, "tactics remembered by older activists were passed onto new ones, from camp maintenance, section 6's, tree climbing and walkways, fighting the legal process, through to the simple joys of stopping work by getting in the way of machinery" (Plows, 2006, p. 13). The emergence of the anti-capitalist/global justice movement (from 1999 onwards) meant the "tactical repertoire of the environmental movement was expanded in the 1990s by alliances forged in the course of the roads protests, and direct-mobilization was legitimized by their apparent success" (Rootes and Saunders, 2007, p. 131). Thus, Adopt-a-Resident is a scheme that combines the outreach, multiplicity of movements (anarchist, ecoActivists, and local residents) and applied repertoires of protest, which in a heterotopic context of are liminoid practices, as ways to share and build the space of the protest

By creating a space at the site of contestation "allows for identification between an uncertain place and marginal and uncertain identities, eclectic, shifting and ambivalent in composition to develop" (Hetherington, 1996, p. 43), held together both by ritualized access and by fostering a sense of community through outreach. The site of contestation becomes the "symbolic sites of Otherness ... through the rituals of festivals, allows an identity to be fully expressed" (Hetherington, 1996, p. 42). An identity is expressed by bringing together "a collection of unusual themes (or discursive statements), and give them a unity of meaning through the production of a space, that acts symbolically as a site for the performance of an alternate mode of social ordering" (Hetherington, 1996, p. 38). Barry (2001) observes that

> the roads protest, along with some other recent forms of direct mobilization, involved a dispersion of politics. On the one hand, political activity was spatially dispersed (...) by developing an inventive form of demonstration, it was possible, however imperfectly and momentarily, to reveal something, which would have been otherwise unknown to others. (2001, pp. 192–193)

The act of protest, the spectacle and "image event" (Deluca, 1999), rely on a series of strategies and tactics built on the evolution and determinant functions of ecoActivism. Within each mobilization, a liminoid (as opposed to liminal) rituals works to define the space (heterotopias). A Liminal ritual, put simply, can be a rite of passage, or a marker that indicates a change in individual status. Liminal rituals are associated with ceremonial practices such as marriage, christening, and funeral and are often ascribed by social order (Hetherington, 1996). Liminoid rituals, on the other hand, are "achieved rather than ascribed … weaker as sources of social integration … and created out of spaces during particular events" (Hetherington, 1997, p. 32). These liminoid spaces are associated with "political protests that have a strong carnivalesque element" (Hetherington, 1996, p. 43). They form the basis of the practice of protest. As tactics and repertoires of protest emerge out of the space of particular camps, a toolbox of tactics is created that forms the foundation of both the heterotopia and the movement. A liminoid example would be the "consensus decision making" practice that occurs at all gatherings, meetings, and mobilizations. Consensus decision-making is a process that excludes neither anyone nor a winner or a loser outcome. BeautifulTrouble.org Toolbox for Revolution describes the process as;

> Consensus decision making is an egalitarian and inclusive method of reaching agreement based on the active participation and consent of group members in collectively reaching a decision. Consensus decision making focuses as much on the underlying processes and values as the decision itself. The word consensus has its roots in the Latin word *consentire*, meaning, to experience or feel together.[2]

Consensus decision making is conducted through a series of hand signals The five main signals are (1) the blocking of a proposal (a fist); (2) a request to make a direct response (both index fingers raised); (3) a point of order or clarity (single index finger raised); (4) a technical point (a two-handed "T" shape); and (5) general agreement, which is indicated with the waving of both hands (Camp for Climate Mobilization Handbook, 2009, p. 10).

EcoActivism, Space, and Sites of Contestation

Protesting at the site of contestation not only shifts the media and political hierarchy, but also alters power relations between activists, the media, and political discourse. In *Manufactured Vulnerability: Eco-Activist Tactics in Britain* (1999) Doherty identifies three factors to have "influenced the development of tactics by protesters" (1999, p. 9). These are (1) situation (site

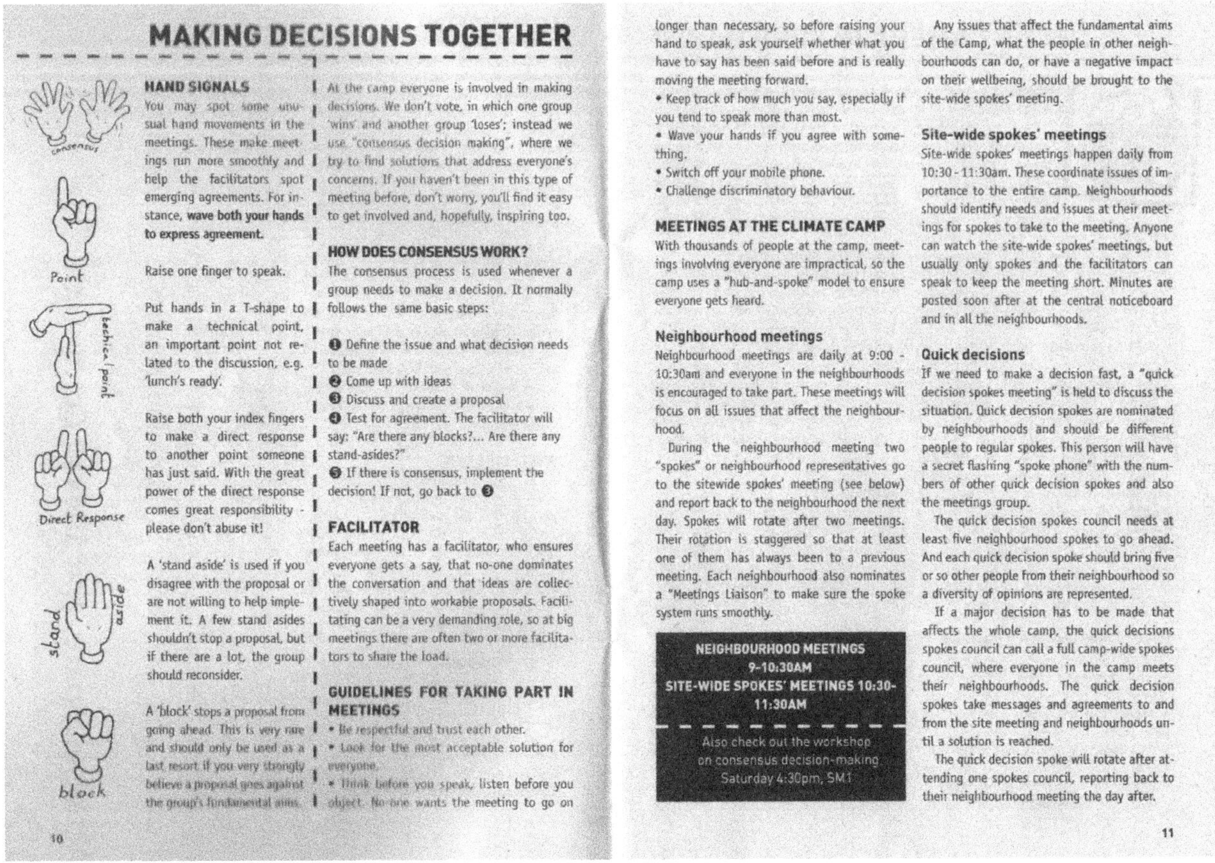

MAKING DECISIONS TOGETHER

consensus

HAND SIGNALS

You may spot some unusual hand movements in the meetings. These make meetings run more smoothly and help the facilitators spot emerging agreements. For instance, **wave both your hands to express agreement.**

Point

Raise one finger to speak.

techical point

Put hands in a T-shape to make a technical point, an important point not related to the discussion, e.g. 'lunch's ready'.

Direct Response

Raise both your index fingers to make a direct response to another point someone has just said. With the great power of the direct response comes great responsibility - please don't abuse it!

stand aside

A 'stand aside' is used if you disagree with the proposal or are not willing to help implement it. A few stand asides shouldn't stop a proposal, but if there are a lot, the group should reconsider.

block

A 'block' stops a proposal from going ahead. This is very rare and should only be used as a last resort if you very strongly believe a proposal goes against the group's fundamental aims.

At the camp everyone is involved in making decisions. We don't vote, in which one group 'wins' and another group 'loses'; instead we use "consensus decision making", where we try to find solutions that address everyone's concerns. If you haven't been in this type of meeting before, don't worry, you'll find it easy to get involved and, hopefully, inspiring too.

HOW DOES CONSENSUS WORK?

The consensus process is used whenever a group needs to make a decision. It normally follows the same basic steps:

1. Define the issue and what decision needs to be made
2. Come up with ideas
3. Discuss and create a proposal
4. Test for agreement. The facilitator will say: "Are there any blocks?... Are there any stand-asides?"
5. If there is consensus, implement the decision! If not, go back to 3

FACILITATOR

Each meeting has a facilitator, who ensures everyone gets a say, that no-one dominates the conversation and that ideas are collectively shaped into workable proposals. Facilitating can be a very demanding role, so at big meetings there are often two or more facilitators to share the load.

GUIDELINES FOR TAKING PART IN MEETINGS

- Be respectful and trust each other.
- Look for the most acceptable solution for everyone.
- Think before you speak, listen before you object. No-one wants the meeting to go on longer than necessary, so before raising your hand to speak, ask yourself whether what you have to say has been said before and is really moving the meeting forward.
- Keep track of how much you say, especially if you tend to speak more than most.
- Wave your hands if you agree with something.
- Switch off your mobile phone.
- Challenge discriminatory behaviour.

10

MEETINGS AT THE CLIMATE CAMP

With thousands of people at the camp, meetings involving everyone are impractical, so the camp uses a "hub-and-spoke" model to ensure everyone gets heard.

Neighbourhood meetings

Neighbourhood meetings are daily at 9:00 - 10:30am and everyone in the neighbourhoods is encouraged to take part. These meetings will focus on all issues that affect the neighbourhood.

During the neighbourhood meeting two "spokes" or neighbourhood representatives go to the sitewide spokes' meeting (see below) and report back to the neighbourhood the next day. Spokes will rotate after two meetings. Their rotation is staggered so that at least one of them has always been to a previous meeting. Each neighbourhood also nominates a "Meetings Liaison" to make sure the spoke system runs smoothly.

NEIGHBOURHOOD MEETINGS
9-10:30AM
SITE-WIDE SPOKES' MEETINGS 10:30-11:30AM

Also check out the workshop on consensus decision-making Saturday 4:30pm, SM1

Any issues that affect the fundamental aims of the Camp, what the people in other neighbourhoods can do, or have a negative impact on their wellbeing, should be brought to the site-wide spokes' meeting.

Site-wide spokes' meetings

Site-wide spokes' meetings happen daily from 10:30 - 11:30am. These coordinate issues of importance to the entire camp. Neighbourhoods should identify needs and issues at their meetings for spokes to take to the meeting. Anyone can watch the site-wide spokes' meetings, but usually only spokes and the facilitators can speak to keep the meeting short. Minutes are posted soon after at the central noticeboard and in all the neighbourhoods.

Quick decisions

If we need to make a decision fast, a "quick decision spokes meeting" is held to discuss the situation. Quick decision spokes are nominated by neighbourhoods and should be different people to regular spokes. This person will have a secret flashing "spoke phone" with the numbers of other quick decision spokes and also the meetings group.

The quick decision spokes council needs at least five neighbourhood spokes to go ahead. And each quick decision spoke should bring five or so other people from their neighbourhood so a diversity of opinions are represented.

If a major decision has to be made that affects the whole camp, the quick decisions spokes council can call a full camp-wide spokes council, where everyone in the camp meets their neighbourhoods. The quick decision spokes take messages and agreements to and from the site meeting and neighbourhoods until a solution is reached.

The quick decision spoke will rotate after attending one spokes council, reporting back to their neighbourhood meeting the day after.

11

Figure 7.1. Consensus decision making: through hand signals, from Climate Camp Handbook (2009).

of contestation as this work defines it)—the terrain and the relationship with the local residents; (2) values and adopting a tribal identity (Liminal); and (3) the ability "of opponents to adapt to earlier tactics" (Doherty, 1999, p. 9). Many of the road protests were situated at places of outstanding natural beauty, wildlife habitats, or, in the case of the Claremont Road, Leytonstone, east London, understood in terms of "social destruction of an east London community, bisected by a road intended to bring commuters from outside London" (Doherty, 1999, p. 9). Once the space is defined, within that space comes the infrastructure of a camp often with walkways, treetop walks, assembly rooms, etc. Treetop homes and walkways combined with tunnels were very common at protest camps in the 1990s. Activists tended to live in either trees or tunnels, each working as a demarcation of the space, without the necessity of a physical fence or wall.

Sites of contestation act as "obligatory points of passage within the network of social spaces that should be viewed as successful heterotopic sites" (Hetherington, 2001, p. 52). Each protest exists as a heterotopia drawing from specific historical tactics so that "those who make these places their own attempt to take on attributes of uncertainty and ambivalence presented by the alternate orderings of such places when performing an ordering of their identities in liminoid ritual practices" (Hetherington, 1996, pp. 38–39). Protest at a site of contestation, create a "counter-conduct that often inverts and subverts "normal social identities and codes of conduct" (Death, 2010, p. 241). All relationships are either invert and subvert not to "exist in the order of things, but in the ordering of things" (Foucault, 1996, p. 38). Thus heterotopic protest camp re-orders the relationship between journalists and activists by drawing journalists to the site of contestation, away from the media focused cities—places of political and economic power, such as government institutions and financial houses.

When direct action occurs away from the "centres of political authority, but in a diverse set of sites" the sites themselves become "places of political activity" (Barry, 2001, p. 192). Many of the camps in the Greater London and southern England region, were "only an hour's drive from London and therefore easily reached by journalists" (Barry, 2001, p. 180). The geographical locale of some protest to metropolitan areas had an impact on the number of journalists attending. A mobilization at the A30 protest near Honiton, Devon (in 1997), over 150 miles from London received less media coverage, with most national newspapers taking reports from the Press Association. Only the *Guardian* sent a journalist down to the site (Barry, 2001, p. 258). Hence, the protest becomes a novelty for journalists, and an inversion of control over media and political discourse swings in favour of the protesters controlling the space. The anti-nuclear, roads protest and later global justice protest all brought to public attention "something that would have otherwise been unknown to others" (Barry, 2001, p. 193), because, prior to the Twyford Down protest, "there was no national roads protest campaign at grassroots level" (Connelly and Smith, 1999, p. 99). Drawing journalists away from the media eye in effect subverts the "symbolic hierarchy of the media frame" (Couldry, 2000, p. 163) shifting power and control of information between protesters and journalist.

Once a protest is over, and removes itself from that space any power gained through counter-conduct is lost. It is within a heterotopia that "revolts of conduct" (Foucault, 2007, p. 194) (when conducts are linked to other conflicts) exist. Protest creates a form of resistance to "power as conducting" through the heterotopia of the camp. A challenge to power through days of resistance (direct mobilization, civil disobedience, site of contestation) and revolts of conducts, to offer an "alternative" (heterotopia) is only successful when the focus and space is in line with the objectivities. Otherwise, a revolt of conducts is "merely dealing with the same phenomena in reverse, from the negative or reactive side" (Foucault, 2007, p. 195), and found in the relation inside a protest with those outside. Nevertheless, what happens when the objective of the space is lost? Friction arises between within the protest from those inside (working groups) and those outside (day or short stay visitors).

Insiders and Outsiders: When a Heterotopic Fails

Feigenbaum *et al.* (2013), summarizes the inside/outside dichotomy succinctly, as "The dynamics of a camp may create insider and outsider dichotomies between different participants, depending on their level of involvement, political backgrounds and other experiences and notions of identity. But these internal divisions and conflicts are also important in understanding protest camps as alternative worlds and places of radical democratic experiences" (2013, p. 2020)

Internal and external modes of action differ in how they view the world outside of the heterotopic space (Roseneil, 1995). An internal mode of action is how individuals act together to constitute a collective, regardless of contrasting ideologies; but there are outward representations through external modes of action—the ways they (social movements) confront the outside world and their political opponents (Offe, 1987 cited in Roseneil, 1995, p. 71). Howarth (2006) observes that "whilst the inside can be constituted through excluding or demonizing the outside (an enemy to be demonized or a state of anarchy to be feared) the outside is not necessarily another whose otherness threatens to subvert or overflow from the inside" (Howarth, 2006, p. 119).

Moreover, governments and protesters depend on an inversion/ subversion as a counter-conduct for a protest to lose traction. Death (2010), "just as government depends upon the creation of governable subjects, such as the liberal citizen, the infirm, the delinquent, the poor, the dangerous and the terrorist, counter-conducts subvert, reproduce, and invert these categories" (2010, p. 32). Resistance to governmentality comes from inverting the "conduct" through "a counter-conducts approach focuses on practices and mentalities of resistance, rather than movements, and also seeks to show how power and resistance, government and dissent, are mutually constitutive" (Death, 2010, p. 240). Activists place themselves outside of society and inside heterotopias as space to practise "mentalities of resistance". They can never remain "outside" as the objective is "mutually constitutive" (Death, 2010, p. 240)

Protest collective identities as agents of change are, "contingent upon their relation to other identities [and] social identities" (Howarth and Griggs, 1998, pp. 55–56). The outcome is the drawing of "boundaries 'insiders' and 'outsiders' [that] requires the construction of 'others' or 'scapegoats' that are presented as blocking the full constitution of an agent's self-identity" (Howarth and Griggs, 1998, pp. 55–56). Governments and protesters depend on invert or subverting Foucault's notion of conduct—"just as government depends upon the creation of governable subjects, such as the liberal citizen, the infirm, the delinquent, the poor, the dangerous and the terrorist, counter-conducts subvert, reproduce, and invert these categories" (Death, 2010, p. 32). To take it a step further, we argue that the repertoires of protest exist within a certain space, and once they remove themselves from that space, they lose any chance of challenging governmentality. It is within a heterotopia that these "revolts of conduct" (Foucault, 2007, p. 194) (when conducts are linked to other conflicts) exist. Environmental activists create a form of resistance to "power as conducting" through the heterotopia of the camp. EcoActivists resistance of green governmentality (that is top-down, economic-led way of conducting individuals) to address climate change in favour of another form of conduct—horizontal, consensus, deep ecology networks. They challenge the power of conducting through their own forms of resistance (direct action, civil disobedience, site of contestation) to offer an alternative (heterotopia) way to be conducted. This research concurs with Foucault's argument that the revolt of conducts is not "merely dealing with the same phenomena in reverse, from the negative or reactive side" (Foucault, 2007, p. 195).

The breakdown of space both from the inside of the camp, and the outside/ inside space of the camp became apparent over five climate camps (2005–2010) because of the openness (the fifth stage of heterotopia). The climate camp's media teams struggled to reach a balance, and far from regulating access for those attempting to enter from the outside (journalists, and police), the entry requirements applied equally strictly to those on the inside. At the same time, openness has provided some activists with the opportunity to (1) engage in movement building; (2) alter the camp's relationship with journalists, thus removing any autonomy; and (3) in the case of Blackheath, a loss of focus in a move away from protesting at the site of contestation. As VisionOnTv's HC says,

> the use of physical space at climate camp is interesting in that the radical media and the traditional media team. At the first one, they pretty much shared the same space, it was pretty low key; in the second one they shared the same space, but I think they might have moved to the tent next door—the traditional media team; the third one, they shared the same space for the first two days and then I think they separated; and the fourth one they just didn't see each other at all." (interview with Hamish, 2011)

Such dislocation is common within heterotopias, because of a propensity to be open to manipulation and alternative directions (the fifth principle of a heterotopia). A heterotopia "always presupposes a system of opening and closing that both isolates [it] and makes [it] penetrable" (Foucault, 1986, p. 26). Saunders and Price (2009) observe that there lies a tension which "will prove difficult to resolve; between those who believe states and corporations alike will end up hijacking the movement with promises of false, profit-orientated solutions, and those whom Monbiot accuses of seeking to "create an anarchist utopia and to use climate change as a way to achieve it" (Monbiot, 2008, cited in Saunders and Price, 2009, p. 120).

The strongest divisions emerge between the older and younger-stalwart, novices and the radical flank of anarchists, who rejected total engagement with journalists at the camp. The radical flank as the role of "radical groups [who] pose a threat to the interests of the State or to other external actors by advocating extreme and politically unpalatable goals, and/or by pursuing those goals using transgressively (often violent) methods" (Gupta, 2005, p. 5). Saunders and Price (2009) note, "within the [climate] camp, the main tension is between the ideal anti-authoritarian self and those more sympathetic to corporate—or state led solutions" (Saunders and Price, 2009, p. 118).

The working groups attempts to engage with social and alternative media had only got the publics' attention so far, and was open to surveillance; relying on the traditional media also became untenable (see Chapters Five and Six), and engaging with a combination of social media, alternative and gate keeping practices caused divisions both inside the protest and outside. Activists have come to realize that engaging with mainstream media brings a new set of problems, and the creation of a hierarchy of top-down politics that begins to speak on behalf of the camp. The movement is no longer a revolt of conducts or counter-conduct, or even a heterotopic space. As Steve (2011) comments the problems began

> divisions grew as a direct result of the media and everybody wanted to get involved, including the mainstream NGOs. When we actually got to the camp, it was a media scrum, you know, the whole street was full of media and we had volunteers from NGOs in their personal capacities who did media for those NGOs who came and helped us. (interview with Steve, 2011)

By opening up the camp, there was friction between those in the camp, and working groups, and the day or short stay outsiders. Steve adds that

> That really did piss off a lot of people who'd put a lot of work into it. Because, it was as John said to stop a third runway and make sure that didn't happen, but it was also to stop it by doing what we do with our aims and objectives for the Camp, and that's when it got diluted and that's when the anti-capitalist stance of the camp. (interview with Steve, 2011)

There was a sense outsiders had little idea of the reason for the camp, and the tension caused problems both with the NGO and day visitors, and within the protest, as people felt the camp was

> diluted by these extra people, and it wasn't their agenda at all but they didn't care that they were joining a movement that had an agenda, they didn't care about that, so that's when those divisions started to occur. (Interview with Steve, 2011)

The press centres, media training, guidance documentation, citizen journalism, open hours, and rules of the camp abandons some of the liminoid social practices. Visually, there are tripods and workshops on repertoires of protest, but the once quasi-structure is formalized for movement building, autonomy is replaced by middle class activists wearing the branding and marketing products of the climate camp. In addition to the inside/outside friction, there was also internal problems with changes to liminal practices and increasing police surveillance.

Fences and Fear: Surveillance and Space

The climate camp infrastructure was supplied by "the TAT Collective stored and delivered tents, marquees and kitchen supplies to protest camps around the UK throughout the 2000s, with recipients including Climate Camps and NoBorders camps" (Feigenbaum *et al.*, 2013, p. 63). TAT were based out of the Leeds EarthFirst! and since 2003 had "in the region of £100,000 has been spent on equipment but with little to show for it". They approached Climate Camp in 2008, to propose they become a group that oversees the structure and working group set up of the climate camp and any future camps on the basis that "a high degree of burn out and exhaustion among those individuals involved in the structure side of putting on large-scale events such as the climate camp. Many of those involved in the ATC (TAT) have been through this mill and recognize how the nature of these events, including pressures from the State, are taking a heavy toll people. The ATC (TAT) will address this by providing a structured approach which minimizes stress and adapts better to the often last minute nature of actually putting events on".[3]

A perimeter fence was built around the entire camp. Fences are symbolic of struggles acting as a determinant function as a symbolic barrier for a movement. Fences are often the first barrier to direct action, and to remove a barrier – with the cutting of fences to gain entry to coal-fired power station, airport runways, etc are liminoid practices. At Blackheath the camp put the fence up, only broken at a controlled point (the entrance) monitored by "gate crews and comms team" (Climate Camp Handbook, 2009). Although the fences were said to protect the camp from the police and outsider interventions, another interpretation of fences could be a sense of the activists conforming by controlling access to the site, and, as Foucault notes, once incarcerated and observed; the "convicts" begin to "behave" (Foucault, 1991, p. 14).

Slowly, the camps cease to be heterotopic as a place of "counter-conduct" (Foucault, 2007, p. 75), nor is it a "movement whose objective is a different form of conduct" (Foucault, 2007, p. 194) looking to be governed—"not to be governed thusly, like that, by these people, at this price?" (Foucault, 2007, p. 75). Instead, it is increasingly looking to "capture the close interrelationship between protest and the forms of government they oppose" (Death, 2010, p. 235). The 2009 Climate Camp was the first away from the site of contestation at Blackheath, South East London. The camp was on common land approximately six miles from its target, Canary Wharf, the financial heart of London.

At the Blackheath climate camp there was too much emphasis on media gatekeeping, a disengagement with outreach, and the shift away from the camps ideology, all enables the fifth principle to occur. Once the space ceases to be an alternative, a heterotopia, it brings to light internal divisions, running beneath the common cause and essentially altered the function of the space. What emerges from the above is the likelihood of division and dissonance, which seems to undermine the plurality and symbiotic heterogeneity inherent in Foucault's definition.

Ultimately, Blackheath was the last large scale climate camp, and a year later when the camps stopped the media tent send out a press release calling the event a metamorphosis of the movement. Metamorphosis is an interesting term, and the beginnings of a shift away from heterotopias, and towards movement building. A series of elements led to the spilt, from the formation of a hierarchy (whether consciously or subconsciously), to generational differences, a lack of historical understanding of the movement and the removal of politics, a point not missed on traditional journalists, means that the space of the camp is no longer a heterotopia.

Foucault's fifth and third stages of a heterotopia are exposed. Stage five—openness—sees power in the camp shifts from collective to a handful of individuals who have been able to take leadership roles because there is openness in the heterotopia. The protest is no longer a unifying space that, regardless of external factors, is united though a common objective, such as economic solutions to the climate debate, a coal seam gas site, an airport expansion, a new supermarket, anywhere that ecological damage is being done, or decisions are made that may lead to ecological damage and climate injustice. The space that was once able to hold several incompatible sites, can no longer sustain the space as individuals come to dominate.

Now through media discourse the camp is mirroring eco-governance selling merchandise and a concept as environmental solutions. Engaging with traditional journalism on a normative scale of press passes, codes of conducts, and gatekeeping to dominate the order of discourse, invites normative behaviour into the movement. The media coverage bring support from those less able understanding of liminoid practices and the heterotopias is no longer outside, but within environmental governance as another consumerist object, a festival, a brand. Activists are no longer "outside", but "inside" society and environmental discourse. Once the movement moves out of the heterotopia, it no longer challenges the dominant discourse. The movement must retain its heterotopic spaces to challenge and survive.

> There was a tension within the media team, therefore, because what was interesting is, Climate Camp itself then became the brand. Rather than it being a No-Brand Space, Climate Camp became the brand but we were doing it both as Climate Camp TV and as VisionOnTv. Now, because we do have a life outside Climate Camp, which is important as well, and um, um, so, so, so there was this kind of tension. I remember we put up a banner, and, and, and someone came and took it down and replaced it with a Climate Camp banner. (interview with Richard from VisionOnTv, 2011)

Ultimately, the end of climate camp movement was a collective decision to move on from a period, to refresh and signpost with other movements on the ground, as one activists says, "the era between 2006–2011 was an incredible period of time. Copenhagen 2009 was a turning point, and building strength on an international level, that's so critically important if you're looking at the impacts of the West and climate change" (interview with Jon 2014). Undercover police exposure, a greater global acceptance of climate change, and climate change issues taking less prominence on the news agenda meant that the end of the Climate Camp movement and spilt was more about understanding where society was at the time, and "things had to move on in terms of political maturity" (interview with activist Anna, 2014). Shifts in the political discourse a greater increase to the End Austerity movements, on "poverty, racial justice, police brutality, things needed to move on to trigger a resonance for thing that was needed … there was a healthy difference of opinion, and police infiltration added to the political immaturity" of the camps, and the decision to end the annual event" (interview with activists Anna, 2015).

Political immaturity could indicate that protest camps are yet to develop into heterotopias; and are instead as Feigenbaum *et al.* (2013) suggest a utopian politics wrapped in a radical alternative. Feigenbaum *et al.* (2013) place the utopian idea which definitely has merit and would be certainly true in the earlier days, but today, with the police surveillance, the burn out risk, the lessons learnt from determinate functions, ecoActivist are realizing that mobilizations can no longer be hit and run located in the here and now, but need to be part of sequential process in political capacity building.

Conclusion

Capacity building within the space of protest could give the strongest indication yet of a longevity within ecoActivism. As the next chapter will show, movement building is now front and centre in the future of the movement. A loss of focus, limited understanding of protest objective, media teams, tents and gate keeping within the movement caused tensions and dislocation within the particular protest cycle between a loss of autonomy, individuals seeking greater power, As the Reclaim the Streets were prominent in the 1990s, the climate camps cessation was due to many things, and ultimately many went onto form Reclaim the Power, the anti-fracking movement, plan for the CoP meetings or focus on wider social and racial justice issues.

The concept of heterotopia provides a useful way to think about the spaces in, and through which the ecoActivist movement contests neoliberal approaches to the climate debate. Various liminoid practices such as direct action, consensus decision making and outreach provide a media spectacle whilst inverting the traditional relationships between activists and journalists. Protest strength is in the space it occupies. How protesters and activists define an alternative space defines the strength of a challenges to neoliberal approach adopted through green governmentality. Despite legislative measures to criminalize the protest camps (as heterotopia) the liminoid practice developed from generation to generation enabled radical environmental activists to challenge environmental discourse.

The media enables activists to have a media presence, promote some citizen journalism and provide some access to turn liminoid practices into new strategies. The internet provides a platform on which to turn protest from the local to the global, as the rhizomatic style echoes the quasi-organizational structures of the environmental activists' movement. Technological developments enhance and aid the creation of heterotopias, such as the swoop/swarm. However, once activists begin to engage with the media within the heterotopia, and the media tent moves from outside to inside the protest camps, the media are no longer outside of the camps, now they are inside the camp with the activists. This can be interpreted as a welcoming in, and acceptance, agreement and conformity by the activists towards a green capitalism, and environmental solutions through economics. In inviting the journalists, they are inviting normative behaviour into the movement. Activists are no longer "outside", but "inside" society and environmental discourse. The movement must retain its heterotopic spaces to challenge and survive.

What emerges from this moment/shift, in the case of the protest camps is a shift from the ideological to the political. No longer does the action involve resisting but instead a redefining their mode of engagement with the outside to align and re-exam their own views rather than those of the outside. We further claim that such an act, closely bound as it is to space will re-signify the camps as heterotopias whose functions change with times. A truly alternative form of engagement with the media might arise in their space. As Graeber (2002, 2009) notes "forms of democratic process that allow initiatives to rise from below and attain maximum effective solidarity, without stifling dissenting voices, creating leadership positions or compelling anyone to do anything which they have not freely agreed to do' then as a movement they also have an opportunity to challenging the current media discourse by reinventing journalistic practices, not mirroring or repeating current long established journalism forms" (2002, p. 71). Where next for ecoActivism? In the next chapter will examine the various strategies that combine the lessons learnt to look to the global stage and the role of activism in maintaining pressure on politicians to enact the Paris Accord and look to the future.

Notes

1. See http://nomadism.org/pdf/taz.pdf for a longer explanation and access to the Bey paper.
2. See http//beautifultrouble.org/principle/consensus-is-a-means-not-an-end/ for further details
3. http://library.uniteddiversity.coop/Money_and_Economics/activist_tat_collective_proposal.doc.

References

Barry, A. (2001). *Political Machines: Governing a Technological Society*, London: Continuum International Publishing.

Bey, H. (1991). *T. A. Z.: The Temporary Autonomous Zone, Ontological Anarchy, Poetic Terrorism*, Brooklyn, NY: Autonomedia. Available at http://hermetic.com/bey/taz_cont.html. (Accessed June 2004).

Connelly, J., & Smith, G. (1999). Politics and the Environment: From Theory to Practice. London, New York, NY: Routledge.

Couldry, N. (2000) *The Place of Media Power: Pilgrims and Witnesses of the Media Age*, London: Routledge

Death, C. (2010). "Counter-Conducts: A Foucauldian Analytics of Protest", *Social Movement Studies*, 9(3), pp. 235–251.

DeLuca, K. (1999). *Image Politics: The New Rhetoric of Environmental Activism*, New York, NY: Guilford Press.

Doherty, B. (1999). "Manufactured Vulnerability: Eco-Activist Tactics in Britain", *Mobilization*, 4(1), pp. 75–89.

Downing, J. (2003). "The Independent Media Centre Movement and the Anarchist Social Tradition", in N. Couldry (ed.), *Contesting Media Power: Alternative Media in a Networked World*, Maryland, MD: Rowman and Littlefield, pp. 243–259.

Feigenbaum, A., Frenzel, F, & McCurdy, P,. (2013). *Protest camps*, London, Zed Books.

Friedman, L. T. (2014). "The Square People, Part 1" and "The Square People, Part 2", *New York Times*, 13 and 17 May 2014.

Foucault, M. (1986). *The Care of the Self: The History of Sexuality*. Vol. 3. Translated by Robert Hurley, New York, NY: Random House.

(1991). "Governmentality". In: Burchell *et al.* (eds.) *The Foucault Effect*, Hemel Hempstead: Harvester Wheatsheaf, pp. 87–104.

(2007). *Security, Territory, Population: Lecturers at the College De France, 1977–1978*, New York, NY: Picador, pp. 333–363.

Graeber, D. (2002). "The New Anarchists", *New Left Review*, (13). Available at http://newleftreview.org/II/13/david-graeber-the-new-anarchists (accessed December 2012).

Graeber, D. (2009). *Direct Action: An Ethnography*, Edinburgh: AK Press.

Gupta, D. (2005). "Radical Flank Effects: The Effect of Radical-Moderate Splits in Regional Nationalist Movements", Prepared for the Conference of Europeanists, Chicago, March 14–16, 2002. Available at http://falcon.arts.cornell.edu/sgt2/PSCP/documents/RFEgupta.pdf

Hetherington, K. (1996). "Identity formation, Space and Social Centrality", *Theory, Culture and Society*, 13(4), November 1996, pp. 33–53.

Hetherington, K. (1997). *Badlands of Modernity, Heterotopia and Social Ordering*, London: Routledge.

Hetherington, K. (2001). "Expressions of Identity: Space, Performance, Politics", *European Journal of Cultural Studies*, 4(1), pp. 109–112.

Howarth, D. (2006). "Space, Subjectivity and Politics", *Alternatives: Global, Local, Political*, April 31, pp. 105–134.

Howarth, D., and Griggs, S. (1998). *Vegans and Volvos, Problematizing and Explaining the Campaign Against Manchester Airport's Second Runway*, Essex Papers in Politics & Government, University of Essex, Dept. of Government.

Krastev, I. (2014). "From Politics to Protest", *Journal of Democracy*, 25(4), pp. 5–19. doi:10.1353/jod.2014.0065

Plows, A. (2006). "Blackwood Roads Protest 2004: An Emerging (Re)Cycle of UK Eco-Action?", *Environmental Politics*, 15(3), pp. 474–484.

Rootes, C., and Saunders, C. (2007). "The Global Justice Movement in Great Britain", in D. D. Porta (ed.), *The Global Justice Movement: Cross-national and Transnational Perspectives*. Boulder, CO: Paradigm Press, pp. 128–156.

Roseneil, S. (1995). *Disarming Patriarchy: Feminism and Political Action at Greenham*, Buckingham: Open University Press.

Rosenwein, B., and Little, L. K. (1998). *Debating the Middle Ages: Issues and Readings*, Malden, MA: Blackwell Publishers, Oxford.

Saunders, C., and Price, S. (2009). "One Person's Eu-Topia, Another's Hell: Climate Camp as a Heterotopia", *Environmental Politics*, 16(5), pp. 117–122.

· 8 ·

POLITICS OF PROTEST

Environmental Activism in a Heated World

"Whenever they cross the redlines we will stop them[1]"

—Climate Games (2015)

Eighteen months out from the twenty first United Nations Conference of the Parties (CoP) meeting in Paris, France, many activists' collectives were already planning events. Six months before CoP21, Reclaim the Power, an overarching group from the UK anti-fracking No Dash for Gas collective held an open meeting to plan for action. Historically, CoP's serve as an opportunity for ecoActivists to take their protest from the local to the global through the megaprotest model.

Gavin and Marshall's (2011) *Mediated climate change in Britain: Scepticism on the web and on television around Copenhagen*, found that generally speaking public understanding of climate change was more confused after CoP15 in 2009 that "Copenhagen suggest that there was an appreciable drop in the number of British people thinking climate change was happening and was largely man-made" (2011, p. 1041). Furthermore, 60% of their survey respondents "had heard something in the news about flaws and weaknesses in the science of climate change" (2011, p. 1041).

This chapter looks at ecoActivism in the context of climate summits with a focus on the CoP21 in Paris. CoP21 was significant for being the first global binding agreement to reduce carbon emissions with the exception of the USA. Activists saw the conference as the most significant climate summit since CoP 15 in Copenhagen six years earlier. This chapter gives a brief overview of UN Climate Change summits, Global environmental governance and the dominance of economics as the common approach to climate change solutions as epitomised by the CoPs. It will then analyse the many activists' attempts to create an alternative space of resistance through artivism and capacity building. The chapter draws on original interviews with activists involved in Reclaim the Power, Dash for Gas, The Hive, End Austerity, Green Party and others to exam the strategies and tactics of the ecoActivists movements that came together for CoP21. Before looking at the history, this chapter will pay respect to those who lost their lives in the terrorist attacks that preceded CoP21.

Friday November 13th, 2015 and the French capital Paris was rocked by a series of terrorist attacks. One hundred and thirty people lost their lives when nine suicide bombers attacked Stade de France, several cafes, restaurants and the Bataclan Theatre. Paris was hosting a series of high profile State meetings, culminating at the end of November with the United Nations Climate Change Conference, the 21st Conference of Parties (CoP21).

Conference of the Parties—A Potted History

The Conference of Parties began in Bonn, Germany (1995) as one of a handful of global meetings with representatives from every country (196) coming together in an attempt to find climate change solutions. Over two weeks, negotiations are made and agreements reached to reduce global greenhouse gas emissions. Climate summits are where the protest movements and the neoliberal advocates come together. Ever since the first global justice/anticapitalism protest in Seattle, USA (1999) there have been mega protest against the political elite, World Trade Organization (WTO) International Monetary Fund (IMF) and multinational corporations.

A series of key global summits produces codes, rules (governance) and forms of conducts for both states and individuals. One of the earliest was the United Nations Conference on the Human Environment in Stockholm (1972). The early conference included a "call from a 'loyalty to the earth … the adoption of global (as opposed to national) responses to environmental problems, and massive changes in over-consumptive lifestyles of the wealthy" (Bernstein, 2003, p. 3). Over the decades subsequent meetings began defining sustainability and environmental discourses, propelled with the ratification of the Montreal Protocol. Montreal Protocol of 1987 saw the global banning of chlorofluorocarbons (CFCs) in white goods and aerosols. CFCs are a major cause of the ozone layer hole (see Chapter Three) and the debate began the shift towards a form of global environmental governance led by the UK, USA and UN.

The next significant milestone the Earth Summit in Rio De Janeiro, Brazil, in 1992 formalised Agenda 21- a key policy heralded as saving the planet. Agenda 21 is "the first draft of the 'Earth Charter', a vision for an environmentally sustainable planet" (Cox, 2010, p. 78). Agenda 21 Charter placed international pressure upon large organizations to adopt an environmental policy. At the Rio Earth Summit (circa 1992), the United Nations Conference on Environment and Development (UNCED) passed a motion to ratify Agenda 21, which "propelled and justified the terms sustainable development into common currency" (Buckingham and Turner, 2008, p. 50). Agenda 21 ratification contributes to the idea for neoliberal solution to climate change, with an objective to "promote market instruments and the integration of environment and development in decision making related action programmes" (Bernstein, 2003, p. 7). Bernstein notes (2003) by the time Rio had occurred there emerged a view which supported the "view that liberalisation in trade and finance is consistent with, and even necessary for, international environmental protection" (2003, p. 4). Increasingly, this view led to the "promotion of global free trade and open markets on the economic side, and the polluter pays principle" (2003, p. 4).

Rio+20 the United Nations (UN) conference on sustainable development marked twenty years since the first United Nations Conference on Environment and Development (UNCED), and ten years since the 2002 World Summit on Sustainable Development (WSSD) in Johannesburg. By 2002 the environment becomes "not only an additional investment opportunity, but also an opportunity for Transnational Corporations (TNCs) and governments to offset some of their environmental wrongdoings" (Finger, 2008, p. 51).

A single tweet sent from an activist collective summed up how much neoliberal economics were embedded in UN and global environmental governance. The tweet from OccupyNeoLibs read "#Rio+20 was a rescue mission, but not for the planet. Its objective was the salvation of the neoliberal model", and sums up the contrasting differences between economics, neoliberal and climate justice modeling to address climate change, the key objective of the United Nations Climate Change policy.

The United Nations climate summits essentially focus on Global Environmental Governance (GEG). GEG provides a global platform to examine climate change, climate justice, and system changes, centred on economic models that favours individuals and corporations. Economic solutions are problematic as they tend to as Bailey, Gouldson, and Newell (2010) says "focus on efficiency and efficacy but have little to say about issues of social justice; they seem to be spatially contingent and to generate patterns of uneven geographical development; and they appear to recreate a neoliberal hegemony" (2010, p. 699).

Nevertheless, at the heart of many UN policies was a strengthening of links between economics and corporate interests. Through a series of newly developed clean energy mechanism such as carbon markets and emissions trading schemes (see Chapter Three), London became home to numerous carbon trading companies. Making billions of dollars a year many global organizations began tapping into emerging green markets. From the New Green Deal (USA), green bank (UK), Green Stimulus packages for business; green economy (United Nations Environment Program UNEP) (Tienhaara, 2014) as a wealth of new green schemes emerges as economic approaches are seen a key to climate change solutions. By the first millennial decade mainstream political parties and global institutions were increasingly co-opting environmentalism into their manifestos and corporate policies. However, many corporations and governments were accused of greenwashing. Greenwashing is when "public relations efforts to portray an organization, activity or product as environmentally friendly" (Beder, 2002, pp. 343–344).

Global Environmental Governance (GEG) is governance that goes beyond the state, but not at the exclusion of the state, focusing on governance between national and international policies. The management of GEC involves multiple actors and mechanisms, emerging out of market-driven, public and private partnerships (PPPs) as the main implementation of global environmental policies such as Agenda 21 (United Nations, 2003) and the Millennium Development Goals (United Nations, 2005). In 2010, the "UN Summit on the Millennium Development Goals concluded with the adoption of a global action plan to achieve the eight anti-poverty goals by their 2015 target date and the announcement of major new commitments for women's and children's health and other initiatives against poverty, hunger and disease" (United Nations, 2011).

The United Nations Commission on Sustainable Development (CSD) into the benefits of Public Private Partnerships (PPPs) says that environmental governance would benefit green issues, because they "provide incentives to the private sector to adopt green criteria" whilst avoiding "politically correct 'add-ons' that mean nothing" (United Nations, 2009). Combined with Kyoto Protocol. which includes market-led ideas to address climate change there emerges a new form of environmental economics dominate by Multinational Corporation and managed through the use of state mechanisms. Governments use the state mechanisms of regulations to beat corporations, institutions and organization into endorsing environmental policies. Yet, at the same time, they allow companies to set their own benchmarks, through voluntary agreements on environmental policies. Moreover, public and large companies have a third trick—moral-suasion or "sermons" (Jordan, Wurzel, and Zito 2003) through eco-labels.

Public-Private Partnerships (PPP) developments were raising concern around levels of accountability and legitimacy of emission trading and carbon markets. Critics of PPPs argue that the division of responsibility between state and capital means a lack of any strong regulation. Without a state-led approach, entrepreneurs and business were, as Thatcher had planned able to work without regulatory boundaries. The state's decision to hand over responsibility to business means the state becomes limited in how much regulation it can impose on business to address the problems of climate change. The emergence of GEG is interest led, not environmental, and there is a lack of effectiveness in measuring the success of any GEG schemes. The hybrid nature of PPPs absolves both state and private enterprises of the responsibility to find solutions to climate change. The state's inability to regulate policy procedure leads to a lack of legitimacy in environmental claims, and limited accountability for the acts of either governments or corporations. This dispersal of reasonability', from hierarchical to networks systems, means that there is a tendency for indigenous or marginalized groups to be excluded or negated in any eco-centric deals (Beder, 2001).

An example of PPP at CoP21 was the exhibition, SolutionCOP21. A public–private partnership, Solutions 21 is a parallel exhibition of the 21 industry sponsors to CoP21. The exhibition's objective was to sell products, services, processes and innovations to fight climate change and its impact. In conjunction with Solutions COP21, the 6th annual Sustainability Innovation Forum seeking business solutions in partnerships between business, governments and public bodies to advance the green economy. Therefore GEG acts as a bridging tool between public and private partnerships. States, at national and European levels, formulate policies and regulations (such as emissions trading systems), regulated through law and city networks. The private sector's role is acted out through markets and networks, voluntary carbon markets and corporate responsibility. Globally, regulation comes from a hybrid of both public and private measures which ensure compliance to MEAs.[2] If summits are the crossroads of GEG, PPP then counter-summits are for activists a space for counter-conduct actions and protest.

Counter-summits are seeking forms of resistance that enables a collective identity as central to social movement theory and literature, and "becoming a collective actor is considered a fundamental task of any social movement" (Price, Saunders, and Olcese, 2014, p. 168). Foucault defines these "points of resistance" as "movements whose objective is a different form of conduct, that is to say, wanting to be conducted differently, by other leaders and other shepherds, towards other objectives and forms of salvation, and through other procedures and methods" (Foucault, 2007, p. 194). Resistance occurs, for Foucault, when power and politics are interlinked through the practice of governmentality, and, as earlier discussion shows, there is a power struggle between activists and party politics over environmental discourse. Politics and power are central to resistance because "everything is political by the nature of things; everything is political by the existence of adversaries. It is a question of saying rather: nothing is political, everything can be politicised, and everything may become political. Politics is more or less than that which is born with resistance to governmentality, the first uprising, the first confrontation" (Foucault, 2007, p. 390). As Death (2010b) notes, "resistance against a pre-determined politics or system" often occurs at times of "heightened conflict across the social system", which produces "intensified interactions between challengers and authorities which can end in reform, repression and sometimes revolution" (Tarrow, 2005, p. 153).

Resistance to governmentality seeks a different way of being ruled that remains within the boundaries of governance—what Foucault calls "counter-attacks, or kinds of reaction (…) are we not dealing with the same phenomena in reverse" (Foucault, 2007, p. 195). There is always "resistance to political discourse and power, and in every epoch, or a crisis of governmentality it is important to ask what forms these counter-conducts take in the current crisis in order to define new modalities of struggle or resistance" (Foucault, 2007, pp. 389–390). By counter-conduct Foucault refers, not to dissent, resist or revolt—all terms he rejects in preference for a term that is as he states;

> in the sense of a struggle against the processes implemented for conducting others; which is why I prefer it to misconduct—which only refers to the passive sense of the word, of behaviour, not conducting oneself properly ... the word 'counter-conduct' enables us to avoid a certain substanstification allowed by the word 'dissidence. (Foucault, 2007, pp. 201–202)

In other words, Foucault's preference for counter-conduct as a term of resistance is akin to acts of dissent, such as civil disobedience, over overt anarchic action. Thus counter-conducts often have the same objective as governmentality, but seek answers and solutions in alternative ways. Resistance comes when those who resist join with similar collectives and movements to form what Foucault terms a "revolts of conduct" (2007, p. 194). Revolts of conduct comes in a new development, a greater push for unity among groups and the Conference of the Parties provides a perfect platform that draws together the last forty years of environmental activism from around the globe.

Paris 2015—Planning to Draw Redlines

Prior to CoP21 numerous ecoActivist working groups across Europe held gatherings to strategize, plan protest and begin to form a media plan. A series of different strategies, collectives, gatherings, social spaces, counter-summits and direct action were discussed and dismissed as plans were revised. One of the first steps was to develop media working groups both at a UK and European level. A European grass roots network, Climate-d 21 formed a working group looking at media, and a websites would be planned and designed as "the first place we're doing any public communication" (Alex activist, 2014)

Discussions over the wording of a unified message between ecoActivist and ENGOs was drafted. Unlike previous CoP meetings, CoP21 quasi-organization structure would see ecoActivists, environmental non-governmental organizations, charities and plethora of different social movement and climate justices groups and organizations coming together under the banner 'system change, not climate change'. From Greenpeace, to 350.org, Reclaim the Power (RtP), Oxfam, Friends of the Earth, and their European counterparts were all deciding on whether to have one message, "a common messages, as a beginning point, so joint common media has a joint common message. It was initially agreed a top line, of something like, leave fossil fuels behind" (interview with Anon, RtP, 2015). Ultimately, the tag line changed to "leave it in the ground".

A key approach was a media friendly name for a planned march on November 29th. Three options were presented 1) Climate rising: The march for climate, justice and jobs; 2) The march for climate, justice and jobs or 3) The march for climate, justice and jobs: People rise up. In the end there were two names, 1) People's Climate March and 2) Global March for Climate which took place around the world, except in Paris. These examples show how activists are now fully conscious that their language will frame the media coverage.

Where to protest and whether to take direct action were also key discussions in the decision making process for the groups. Whether to take action at the CoP21 venue, La Bourget or the Parisian city centre was varied. CoP 21 venue was fifteen miles outside of central Paris in a purpose built hanger at Le Bourget airport. At a RtP planning meeting discussions centred on the merit of direct action at the site of contestation. Should the protest take place either at the conference centre (Bourget, near Paris International airport) a Parisian landmark or offices of an energy company? As one activist explains to a group, to get media coverage meant being strategic about choosing which site of contestation was suitable, he says;

> if there's no clear idea of why go to the Bourget, then don't go for the hell of it, or because it's the obvious point to go, because there has to be some protest outside. It's better to do something tactically different like focus on *Total*, the flag ship oil company in France and France is the host of CoP21. (Anon, RtP, 2015)

During the meeting there was a general agreement that a balance was needed as one activists said, there's a need for "something around the summit site, more iconic, and mass action around it, and at the same time" (RtP general discussion). Another group member added that any strategy needed to include media eye concept and image events, she says, "Then there's the problem if you do decentralize things, and you do stuff in the city, Paris proper. There's not the target and how do you make that into a story. Is there a way to make them both work in some way? We can discuss what action we could do, but is that out of our hands and to some extent will be decided by 350.org and various others and we'll fit, or slot into that" (RtP, 2015).

Before CoP21—Plans and Paris

Direct actions strategies were discussed to keep the focus on decision-making processes. A possible tactic—locking the delegates *into* Le Bourget was considered, but was ultimately determined to be unnecessary—was only "If it's not a good enough deal, there's talk about shutting them all in at the end. It's tactically difficult to shut them in" (Anon, RtP, 2015). Lock-them-in was one idea discussed and rejected over concerns of media misrepresentation. As one person said, "lock them in, feels like, in terms of strategically or what we want out of it, and our story, given what's happened at every other CoP, it could turn a potentially negative things into a positive of people inside and outside are taking matters into their own hand, It could be really powerful", whilst another was worried about antagonizing those inside Le Bourget. Whilst Rob noted a lock them-in strategy could be problematic, noting that

> those on the inside, there are more than 10,000 registers civil society media on the inside, and talking about the potential to lock them in, and those on the outside beyond the riot cops, there's potential for two lines of action. You could also imagine that small island states etc., could be part of that, because once you get to the end, then there's only enough room for the minster to sit behind the table. (RtP, 2015).

There was agreement that direct action and civil disobedience would need to be media friendly, and not necessarily have a set focus. As someone who had just returned from a grassroots meeting in Europe explains,

> We as the UK, don't have the capacity, of whatever we're presenting, to decide our big action. What we do have is knowledge and the possibility to influence and take part in the Climate Games, an inclusive way of taking action. To galvanises action in the UK and Paris (...) so we feel we are all united. (RtP, 201)

Discussions focus on whatever was to galvanize would have to also be media directed. As one returner says, "Direct action groups do something in the airports, symbolic, visual, creative things that look good in on TV and the press where you can't get inside, and give them something. In terms of messaging, it's very difficult to get the key things we want".

Movement Building and Reclaim the Future

CoP21 Mobilization was about two things—movement building and a unified media message. Movement building is the acting in solidarity and forging links with people and groups with common or related interests, including workers and the communities or populations most acutely affected by climate change in Britain and throughout the world, to build a movement with the wisdom, diversity and strength to achieve true ecological and social justice

The first action, BreakFree 2016 was to be a political uprising against economic disparities, protest against living standards, human rights, divisions in wealth (such as the 99% Occupy protest). As part of a movement building strategy, there was a constant iteration to stay on message. Coalition Climate 21 (CC21) media working group, put out suggestions for media talking points and "They are meant to be used by the coalition when speaking to the media, and to be used by members / partners when referring to the coalition / collective process". Talking point for N28/29—On November 28/29 included demands for Climate Justice, to show that the climate struggle is also a struggle about inequality and poverty, about food and energy, about water, jobs and human rights.

CC21 a collaboration of 130 organizations, from civil society, labor unions, religious, human rights, environmental and social movements. CC21 local working group organizes an open space, a social forum type space, possibly similar to the Klima Forum in Copenhagen 2009;[3] the D12 march, and sleep spaces. Avaaz's planned event in—Place de la République filled with shoes to symbolize the footsteps of all. The action went ahead despite La République being one of the sites of the terrorism attack. Those leaving shoes behind was the then UN Secretary General, Ban Ki-Moon with a note reading "as the world gathers in Paris, to stand up for climate action, let us also stand in the shoes of the victims of terrorism, war and persecution … and respond with compassion".[4] A Vatican representative left shoes from Pope Francis. An estimated 10,000 shoes were left as a memorial to the victims of the Paris attacks and climate change.

CC21 coalition press release (25 November 2015) outlining that despite the attacks and the state of emergency "civil society will demonstrate its concern through a variety of creative initiatives. These have been instigated by member organisations of the Coalition Climate 21, which acts as a sounding box for this mobilisation of civil society". Others, such as Avaaz and Alternatiba and their human chain for climate peace an echo of Embrace the Base at Greenham Common Peace Camp (1982); Great Climate Clamour—a similar event to Global Noise Protest a year after the Occupy camps closed, saw protesters come together to make noise, with pots, pans and drums.

In Paris, D12—, was the main day of action and the final day of CoP21. D12 talking points were to be people, not politicians and polluters. The global march was to make it clear to the world that Paris is not the end of the fight for climate justice, but a moment in the continuation of an ever stronger, growing and more unified movement.

Other actions included a Citizen Summit for Climate based in the East Parisian suburb of Montreuil as an alternative summit of workshops, debates, screening, performances and presentations. The summit was exploring concrete alternatives to the current solutions put forward on climate change.[5]

Place to B, in Le Garde De Nord train station was acting as a hub for freelance and alternative journalists outside of the conference. Place to B was reworking of an earlier similar space, Climate Outreach. Climate Outreach is according to its website "Europe's leading climate change communicators, bridging the gap between research and practice and helping to widen engagement across a broader spectrum of society".[6] Elsewhere, a global critical mass ride[7] had Reclaim the Power and other cycling collective riding from London to Paris to join up with the global critical mass.

Spoof advertisements popped up across Paris, and the most common, were the 600 culture jammed advertisements artworks by the artivist collective, Brandilism. Brandilism's concepts are critiques of the "corporate takeover of the CoP21 climate talks". Altered images were installed across Paris in street furniture, owned by advertising corporation, JC Decaux. A global advertising agency JC Decaux own 491,950 advertising panels in 56 countries that generated 45.3% of their annual income in 2014. Brandlisim's civil disobedience was to use the spoofed artwork (Dekeyser, 2015) to "illegally replace existing advertisements with artworks that raise concerns around the sponsorship of CoP21 by large corporations such as Engie (previously GDF Suez), BMW, Dow Chemicals, The Coca-Cola Company and Air France". Brandalism's Joe Elan accuses the 21 sponsors of blatant greenwashing, "we are taking their spaces back because we want to challenge the role advertising plays in promoting unsustainable consumerism. With the Climate talks and their corporate sponsored events, outdoor advertising ensures that those with the most amount of money are able to ensure that their voices get heard above all else." Those behind Brandilism have previous connections with street artist Banksie and provide instructions for opening advertising hoardings.[8] The Paris action was targeting the 21 companies sponsoring CoP21.

State of Emergency and Public Statements

Twenty four-hours after the attacks, the French government implemented a state of emergency, the first of its kind since the Algerian war of independence (1945–1962). Then French President François Holland, announced a nationwide state of emergency. The etat d'urgence[9] saw 1,500 military police on the Parisian streets. Police were given extra powers to prevent people and vehicles from moving in areas fixed by decree, including the cancellation of all planned school trips.

The implementation of curfews saw the banning of large gathering and marches; and police were given extra powers to carry out house searches without a warrant; enforce house arrest; remove any weapons whether legal or illegal, and forbade protests. After 12 days, the state of emergency was extended up to three months, including timeframe of CoP21. Anyone in "Violation of these extraordinary measures can result in a prison term not to exceed two months, a 3,750 Euro fine, or both" (Severson, 2015).[10] Almost two years later the etat d'urgence was lifted and replaced by new counter terrorism laws.

Several protesters were placed under house arrest until the completion of CoP21. *The Guardian* newspaper reporting 24 people were placed under house arrest for planning and organizing large-scale protest. Their movements were restricted to leaving the house three times a day and required to sign a register verifying their whereabouts. Author and journalists Naomi Klein is quoted at the time as saying "Climate summits are not photo opportunities to boost the popularity of politicians … given the stakes of the climate crisis, they are by their nature highly contested. That is democracy, messy as it may be. The French government, under cover of anti-terrorism laws, seems to be trying to avoid this, shamefully banning peaceful demonstrations and using emergency powers to pre-emptively detain key activists" (Neslen, 2015).

Paris—Post Nov 13

In light of the attacks protest group issued press statements on any planned action. Various ENGO and NGOs and ecoActivism collectives put on a united front to underline that the attacks were an inhumane act, but, "The Paris attacks make climate protests more important than ever" (RtP). International Attac a movement begun in the late 1990s as part of the alter-globalization movement—a "global network of people's organisations opposing the ultra-free-market policies being pushed by the governments of the world's richest countries".[11] Global Justice Now in the UK is one of many chapters across the globe that "operates autonomously, but common campaigning".[12] Attac's statement read "We reject in advance any restriction on the right to protest and fight against this decaying world, in favour of the alternatives that peoples of the South and the North put forward together". From November 29 to December 12 in Paris on the occasion of the CoP21 and with our citizens' mobilizations, we will show that another world is possible, necessary and urgent. Climate Game's statement read

> First of all, we want to clearly state our solidarity with all victims of all forms of terror. Machine guns and explosives hurt the same whether in Paris or Beirut, Ankara or Yola, Damascus or Kobane, Baghdad or elsewhere. The hurt feels the same whether it comes from the gun of a jihadist or a police officer, the missiles of a fighter plane or a drone. These attacks must not change the conversation but deepen it. We want to clearly state that our dedication for social and climate justice remains as strong as ever. We are convinced that the geopolitical and economic dynamics that underpin climate chaos are the same as those that feed terrorism. From the oil wars in Iraq to the droughts in Syria caused by ecological collapse, all feed the same inequalities that lead to cycles of violent conflict".[13]

Culture Jammers, Brandalism responsible for over 600 subverted advertising posters, follow a similar tone as Climate Games and 350.org. Their statement reads, "Following the tragic events on 13th November in Paris, the government has chosen to ban the big civil society mobilizations—but big business events can continue. The multinationals responsible for climate change can keep greenwashing their destructive business models, but the communities directly impacted by them are silenced. It's now more important than ever to call out their lies and speak truth to power. We call on people to take to the streets during the CoP21 to confront the fossil fuel industry. We cannot leave the climate talks in the hands of politicians and corporate lobbyists who created this mess in the first place."

REDLINES—A Movement Building Project Beyond Paris

Redlines was a strategy that symbolizes the crossing of a line to the point of no return. Redlines objective set out before the attacks and loss of 130 lives, was set for December 12th 2015 called D12, following the patterns of J18, N30 (1999) and was targeting the final day of the CoP21 meetings. Redlines were the visual meme of the image event, and symbolic line for a sustainable planet. The plan was to surround the conference centre, Paris-Le Bourget site and build a wall of "cobblestones" as a symbol for the redline that must not be crossed in search for climate change solutions.

Tools for action made inflatable cobblestones with dimensions of 1 meter 40, by, 1 meter 40 and 1 meter 40. Inflatables also alter the space and power relations between police and protesters. Just as the protest camps shifts and alters the power relations between journalists and protesters when journalists enter into, the camp as activists control the space of the protest camp, inflatable's work in a similar way. The inflatable subverts the protest space, defuses the power relations, acts as a symbol of creativity and control by protests and autonomous, it's much more difficult to control inanimate objects and at the same time remove the message—the cobblestone, the resistance. E.e.c. artivists, Artúr van Balen and Verena Meyer speaking to blogger, Johanna Rainer (2012) says,

> Inflatable's are like giant puppets. They strike the imagination … People who are smashing windows are easier to control because you can criminalize them; but a symbol of collective creativity is much harder to denounce. And: You cannot give instructions to a gigantic inflatable or puppet. It transforms the situation into a carnivalesque celebration. Inflatables embody the idea of tactical frivolity (Rainer, 2012)

Made from silver insulation material common in the building industry, the inflatable cobblestones were a homage to the *Situationist* movement as a means of subverting the police and any protesters planning on dispersing protest with force. Ideas from the artivist collective eclectic electric collective (e.e.c.), the silver cobblestones are both a "celebration of an object which is both a symbol and a material weapon of anti-authoritarian struggle everywhere. It also aimed to bring new strategies of tactical frivolity into the demonstration".[14] The new strategy, aside from police and protesters to-and fro with a large inflatable, is the image event of a game, a child's toy, that media framing has no choice but to use the "photos and films would be in favour of the protesters. Because it is also about the politics of images."[15]

The *Situationist* movement of 1968 Paris had the idea "Beneath the cobble stones the beach". Beneath the cobble stones the beach is a line from the *Situationists*, a group of students from the Sorbonne University, who on May 1st, 1968 took to the streets of Paris. "Under the cobblestones was a "piece of wall graffiti made in Paris in May '68, which represented the May events as a moment of political and social transformation: if you ripped up the cobblestones the possibility of a transformed social order opened beneath" (Jones and Skaife, 2009, p. 19).

Guy Debord's a key player in the *Situationist* movement wrote in *The Society of the Spectacle*, drawing on the events of Paris, 1968 that the only way to dismantle the hegemony and create a society with our own principles is that we cannot dismantle the master's house without the masters tools. In other words, individuals, social movements, collectives, and subversive groups must use the techniques of the capitalist culture to counter the effects. The radical needs to contest the control over space, production and who is allowed in and out of a society. Whereas Foucault showed that the state took control, for Debord radicals and protesters must take control over the state.

Such images, and image events are one way to "changed the dynamics of public communication and have had a significant effect on the process of social issue construction and public opinion formation" (Delicath and Deluca, 2003, p. 320). Combined with the appropriation of space defined and refined through decades of protest camps, the idea of subverting public space to contest hegemonic positions is one adopted during D12. Drawing on Debord's concepts that the use of the spectacle in public life is a means of contesting cultural hegemony, then the development of the Redlines concept is a way to contest political decision-making processes, in this case, the CoP21 meeting and future meetings within any cultural and public space.

Paris was always about movement building. A Redline strategy is no longer a reactionary movement reacting from one policy announcement to other, but a planned, co-ordinate movements with action planned. Prior to Paris, the anti-coal collective, *Matt Ridley's Conscious* carried out an action at the entrance to Lord Ridley's coal mine. Shotton Surface mine, on the Blagdon Estate, Northumberland, North East England. The estate is owned by the fifth Viscount, Matt Ridley with approximately six million tonnes of coal to be mined from the land. Mining company, Banks Mining has been mining the site since 2008. EcoActivists blockaded the entrance to the mine for eight hours (27 October, 2015). One group d-locked themselves to a digger, whilst another blockade the mine entrance halting business for a day. A small group locked on to each other at the mine entrance, their hands in locked together in long red plastic tubing. The red painted tubing had the words "redlines" written on them. *The Guardian* reports that "Guy Shrubsole, 30, one of the protesters padlocked to the diggers who was later arrested, said the group hoped the direct action would draw attention to Ridley's 'self-interest' in the coal industry, as well as promoting the environmental importance of keeping fossil fuels in the ground in the run-up to UN climate change talks in Paris next month."

> The Red lines are the minimal necessities for a just and liveable planet. More specifically: Sustainable energy transformation / Justice for impacted people / the right

> to food and water. In order to avoid dangerous climate change we know we must keep at least 80% of known fossil fuel reserves in the ground. We must completely decarbonise our economy before 2050, with a peak in emissions by 2020. Keeping these fuels in the ground is our movement's red line. A minimal necessity for a just and liveable planet.[16]

Redlines are a collaboration between environmental NGOs such as 350.org, Redline's objective is not to get into the conference centre but to lie down Redlines around the site and occupy public space with disobedient bodies. There were no plans to break through Police lines but keep a distance with the eyes of the world's media on them and contradicting the French state's desire for a totally controlled and smooth summit

Whilst Redlines were one of many groups coordinating action in 2015 and beyond. ReelNews.org report that "Most of the people who were there as activists went not to affect change, but the main reason, why most people went there was to organise a far more militant action collectively over 2016"(interview with Shaun, 2016). Whilst some people went for the marches and rallies, others were there "not to protest the deal directly, and to talk about how do we change the system", through future protest events, direct action and civil disobedience. Break Free From fossil fuels 2016 (breakfree2016.org)

Reclaim the Power, anti-fracking collective in the UK also took Paris as an opportunity to both highlight corporations power in influencing global negotiations, but also, the Reclaim the Power will use CoP21 as a "platform to build our movement and amplify our tangible demands & key messages to the media and the wider public, as well as corporations and policy makers" (RtP message board, 13 April 2015).

Conclusion

CoP21 began under a cloud of terror and sadness in the wake of 130 murders. Tagged as a game changer for the environment, CoP21 was also to be a new beginning with a legally binding agreements for a universal agreement whose main aim is "to keep a global temperature rise this century well below 2°C and to drive efforts to limit the temperature increase even further to 1.5°C above pre-industrial levels".[17] Economic driven solutions overseen by the United Nations-led NAZCA (Non-State Actor Zone for Climate Action) countries are now legally committed to reducing global carbon emissions. UN Framework Convention on Climate Change (UNFCCC), Executive Secretary

Christiana Figueres is quoted as saying "The recognition of actions by businesses, investors, cities and regions is one of the key outcomes of CoP21. Together with the LPAA, the groundswell of action shows that the world is on an inevitable path toward a properly sustainable, low-carbon world." The next test will be in 2020 when the Paris Accord is reviewed and targets amended. CoP21 wasn't about finding solutions for the protest movements; it was a test ground for movement building.

Earlier mobilizations by Reclaim the Power, No Dash for Gas, Climate Camp were all debates around 'clean coal' and aviation exposes conflicts between different agendas, and the problems and possibilities of maintaining a radical focus on social justice within the environmental movement, as the climate camps "explicitly sought to re-introduce a political space" (Schlembach, 2011, p. 14). The climate summit are capstones drawing together the last forty years of protest actions, finding ways around and a realization that climate summits, direct actions, global environmental governance at local and global levels, neoliberalism are all vying to shine a light on the political space of the climate debate.

Notes

1. https://www.climategames.net/docs/d12.en.pdf
2. For example, the Kyoto Protocol of 1992 or, more recently, the World Summit on Sustainable Development, held in Johannesburg during 2002 and the UN Climate Change Conference in Copenhagen in 2009.
3. Klimaforum09 is part of the movement that is building up from everywhere towards the COP15. Klimaforum09 is initiated by the independent association Civil Society's Climate Forum, which is anchored in a coalition of 21 Danish social and environmental organisations and movements—supported by more than 40 international movements, networks and organisations. The association is open to anyone who wishes to participate in creating Klimaforum09 and can comply with the Platform.
4. https://twitter.com/UN/status/670914109169410048/photo/1?ref_src=twsrc^tfw
5. https://climatejusticeaction.net/en/obj/convergent-spaces-in-paris/
6. http://climateoutreach.org/#
7. http://velorution.org/d5/en/
8. http://www.brandalism.org.uk/bus-stop-guide.pdf
9. http://legifrance.gouv.fr/affichTexte.do?cidTexte=JORFTEXT000031473404
10. https://www.lawfareblog.com/state-emergency-how-paris-attacks-expanded-frances-police-powers
11. http://www.globaljustice.org.uk/attac
12. http://www.globaljustice.org.uk/attac
13. https://www.climategames.net/response-to-the-recent-attacks

14. http://truthisconcrete.org/interviews/you-cannot-give-instructions-to-a-gigantic-inflatable/
15. http://truthisconcrete.org/interviews/you-cannot-give-instructions-to-a-gigantic-inflatable/
16. https://www.climategames.net/docs/d12.en.pdf
17. http://newsroom.unfccc.int/unfccc-newsroom/finale-cop21/

References

Anon. (2015). Comment made by anonymous activists at Reclaim the Power (RtP) meeting, Hive, London.

Bailey, I., Gouldson, A., and Newell, P. (2010). Ecological modernization and the governance of carbon: a critical analysis; July, Centre for Climate Change Economics and Policy Working Paper No. 26 The Governance of Clean Development; Working Paper No. 09, July. Available at http://citeseerx.ist.psu.edu/viewdoc/download?doi=10.1.1.368.7645&rep=rep1&type=pdf

Beder, S. (2001). "Greenwash", in J. Berry and G. Frankland (eds.), *International Encyclopaedia of Environmental Politics*, London: Routledge.

Beder, S. (2002). *Global Spin. The Corporate Assault on Environmentalism*, Glasgow: Green Books.

Bernstein, S. (2003). Liberal environmentalism and global environmental governance, *Global Environmental Politics*, 2(3), pp. 1–16.

Buckingham, S., and Turner, M. (2008). *Understanding Environmental Issues*, London: Sage.

Cox, R. (2010). *Environmental Communication and the Public Sphere*, Thousand Oaks, CA: Sage.

Death, C. (2010). *Governing Sustainable Development: Partnerships, Protests, and Power at the World Summit*, Hoboken, NJ: Taylor & Francis.

Debord, G. (1995). *The Society of the Spectacle*, New York, NY: Zone Books.

Dekeyser, T. (2015). "Why Artists Installed 600 Fake Adverts at CoP21", *The Conversation*. Available at http://theconversation.com/why-artists-installed-600-fake-adverts-at-cop21-51925

Delicath, J. W., and DeLuca, K. M. (2003). "Image Events, the Public Sphere, and Argumentative Practice: The Case of Radical Environmental groups", *Argumentation*, 17(3), pp. 315–333. doi:10.1023/A:1025179019397

Finger, M. (2008). "Which Governance for Sustainable Development? An Organizational and Institutional Perspective", in J. Park, K. Conca, and M. Finger (eds.), *The Crisis of Global Environmental Governance: Towards a New Political Economy of Sustainability*, New York, NY: Taylor Francis, pp. 34–57.

Foucault, M. (2007). *Security, Territory, Population: Lecturers at the College De France, 1977–1978*, New York, NY: Picador, pp. 333–363.

Gavin, N. T., and Marshall, T. (2011). "Mediated Climate Change in Britain: Scepticism on the Web and On Television Around Copenhagen (Report)", *Global Environmental Change*, 21(3), 1035–1044.

Jones, K., and Skaife, S. (2009). "'Under the Cobblestones, the Beach': The Politics and Possibilities of the Art Therapy Large Group", *Psychotherapy and Politics International*, 7(1), pp. 18–27. [Article]: Goldsmiths Research Online.

Jordan, A., Wurzel, R., and Zito, A. (2003). "'New' Environmental Policy Instruments: Patterns and Pathways of Change", *Environmental Politics*, Special Issue, 12(1), pp. 1–25.

Neslen, A. (2015). "Paris Climate Activists Put Under House Arrest Using Emergency Laws". Available at https://www.theguardian.com/environment/2015/nov/27/paris-climate-activists-put-under-house-arrest-using-emergency-laws (accessed January 2016).

Price, Saunders, and Olcese. (2014). "Movements", in C. Death (ed.), *Critical Environmental Politics*, London, New York, NY: Routledge, Taylor & Francis Group.

Tarrow, S. (2005). *The New Transnational Activism*, Cambridge: Cambridge University Press.

United Nations. (2003). *Agenda 21: Earth Summit – The United Nations Programme of Action From Rio*, April, New York, NY: United Nations Publications.

United Nations. (2009). *Good Governance in Public-Private Partnerships*, United Nations Commission on Sustainable Development (CSD), New York, NY: United Nations Publications. Available at http://www.un.org/esa/sustdev/csd/csd16/PF/ presentations/hamilton.pdf

United Nations. (2011). *Millennium Development Goals Indicators. The Official United Nations Site for the MDG Indicators*, updated 2011. Available at http://www.un.org/millenniumgoals/

Rainer, J. (2014). "You cannot give instructions to a gigantic inflatable". Retrieved from URL http://truthisconcrete.org/interviews/you-cannot-give-instructions-to-a-gigantic-inflatable/#more-128

Schlembach, R. (2011). How do radical climate movements negotiate their environmental and their social agendas? A study of debates within the Camp for Climate Action

Sevenson, D. (2015). State of Emergency: How the Paris Attacks Expanded France's Police Powers. Retrieved from https://www.lawfareblog.com/state-emergency-how-paris-attacks-expanded-frances-police-powers

Tienhaara, Kyla. (2014). Varieties of green capitalism: Economy and environment in the wake of the global financial crisis. *Environmental Politics*, 23, 187–204.

· 9 ·

CONCLUSION

Environmental Activism and the Media

> I think it's really easy these days to feel crushed an impotent by society and government. If you can do something that makes you feel involved, makes you feel like you're making a contribution, I think it's the keystone for a much healthier and happier lifestyle.
>
> —Interview with Greenpeace Rainbow Warrior Captain, Peter Willcox (2013)

This final chapter offers a summary of the central arguments made in *Environmental Activism and the Media: the Politics of Protest*. Drawing together the three themes of the book, environment governance, ecoActivism and media discourse the work has looked at relations of power between all three in relation to the politics of protest. Firstly, the work focused on the greening of politics, tactics of law against ecoActivists and the use of political policing to gather data on protests leading to some being labelled domestic extremists. Chapters Five and Six looked at how by becoming a media movement as early adopters of online activism, this tactical media approach helps give a voice to a movement. Chapters Seven and Eight looked at the strength in movements from the space of protest to movement building and reclaiming the future to influence the moveable middle. In a highly politicized and mediatized world, as climate change becomes more and more prevalent within a plethora of political, cultural and economic discourses highlighting key strategies of success (and identifying failure) in representation of environmental activism in the media, this book aims to leave readers with a new knowledge, insights and historical understanding, when embarking on any form of environmental activism and provide academics and activists with the tools to theorize key issues in the politics of protest.

There have been, and continue to be, many discursive moments that shape public perceptions of ecoActivism. From the late 1960s to 1990s, ecoActivists were at times the primary definer of socio environmentalism and global warming issues. Politically, both domestically and internationally has seen economics and environmentalism become immersed in the climate change debate. By the end of the century, the loss of the ecoActivist voice is evident in the traditional news reporting, replaced by multinational corporations and political rhetoric.

Media discourse initially gave activism a voice (boundaries of representation in Fairclough's (1995, 2000) terms. Post-1997 and the incorporation of global justice movement and anti-capitalist rhetoric into environment activists' discourse sees a power shift within environmental discourse. This shift is reflected in the relationship between activists, the state and media discourse. Thus the media supports the state over the collectives, but will occasionally support the individual (as indicated, for example, by the discussion on Swampy).

A green governmentality (Luke, 1999; Oels, 2005) approach saw consecutive governments encourage the public to buy green products, leaving the individual to conduct (conduct of conduct) themselves for the greater good; whilst protest and activism was increasingly labelled a deviant act. The introduction of new laws, the expansion of police powers, increasing funding for undercover policing and surveillance tactics was supported by journalists, whose reporting added to the public perception that ecoActivists were deviant. Globally, the Kyoto Protocal's ratification (1994/2006) opens the way for alternative energy markets, clean energy mechanism, emissions and carbon trading schemes, all indelibly linking climate change solutions to economics. Mid 2000s and the rhizomatic expansion of the internet, matching the horizontalist non-hierarchical social movements took protest from the local to the global linking environmental protest to global justice/anti-capitalist movements via cyber-protesting, social media, and websites. New media and a digitally literate generation led to citizen journalism after realising engagement with traditional journalists is a luxury not to be ignored (Chapter Five). The exposure of undercover policing with PC Mark Kennedy (in 2010), also revealed the predatory nature of state-sponsored surveillance.

The Future Politics of Protest?

Drawing on Luke' (1999) and Oels' (2005) interpretations of eco-governmentality through a green lens, this historical approach provides an examination of how environmental discourse in the media began with a high concentration of coverage on activism. At the same time, there was the fledgling development of environmental governance. As activism was in the media spotlight, governance sat back, and with Margaret Thatcher's Green speech (1988), came the voice of party politics into environmentalism.

The creation of the Department of Environment (DoE) in the early 1970s as an administrative tool that gave power to governments to regulate everyday environmental issues, from town planning to transport systems. New towns and transport routes were designed not as sustainable, but functional moving the urban city sprawls further out into rural areas. Through governmentality and biopower (Foucault, 1971, 1972, 1986, 1991, 2008), it also gave the government power to decide where people would live, how they would travel and what facilities they would live by. These techniques could then be applied to influence individual behaviour through political decision-making processes (biopower), persuasion and conduct of conduct.

Through techniques of self-government (conduct of conduct in Foucauldian terms) rules and regulations (governance) are developed to guide the individual to act in a way that is of benefit to society as a whole. The individual is encouraged to conduct themselves in a way that benefits society before they can conduct others (the conduct of conduct). Moreover, institutions such as the Department of Environment provide the administrative tools through which individuals can conduct themselves. The creation of administrative tools over the individual (biopower) becomes more prevalent in persuading individuals to act in ways believed to be of greater economic benefit to the individual and country. Individuals who challenged or rejected this neoliberal discourse through civil disobedience and direct action are often criminalized for their abnormal behaviour (techniques of dominance). Thus Biopower provided the tools (techniques in Foucauldian terms) to encourage individual behaviour and contain those who reject new neoliberal approach to a whole range of social problems.

Criminalizing elements of Direct action and civil disobedience was largely the result of changes to the law and more power to police units. The Criminal Justice Bill (CJB), later the Serious Organized Crime and Police Act (SOCPA), repealed and replaced with the Police Reform and Social Responsibility Act

(PRASRA) and anti-terrorism measures that includes protest around sites of national security interest, led to what Luke (1999) suggested would recast ecoActivism "as a type of civil disobedience, which endangers national security, expresses unpatriotic sentiments, or embodies treasonous acts" (1999, p. 125). At the same time, more subliminal and subtle legislative measures were used as surveillance techniques on activists' movements. The plethora of new police intelligence units providing information on protest groups, National Domestic Extremism and Disorder Intelligence Unit (2004) (NDEDIU); National Public Order Intelligence Unit (1999) (NPOIU), National Extremism Tactical Coordination Unit (2004) (NETCU), and National Domestic Extremism Team (NDET); Counterterrorism Command (2006), Protection Command and Security Command were often in the role of supporting corporations from large-scale gatherings counter terrorism and security. Each unit was drip feeding intelligence and in some cases being agent provocateurs, to increase the criminalization of protest in the UK, USA, Australian and New Zealand (see Chapter Four).

Funding by the Home Office for the four new units gave the police more powers to use undercover policing practices, and later to access activists emails, databases and mobile phone messages. The rationale for these new surveillance powers was that they were needed as preventative measures against potential acts of violence (as Luke, 1999 predicted). Violence and a fear of violence is epitomized in a certain period conflating activism with terrorism. Lord Pitchford who was leading the inquiry into undercover policing practices in the UK has been replaced by Sir John Mitting. Lord Pitchford stepped down as chair due to ill health. Activists are still waiting the outcome of the inquiries decision whether the matter of people who may have been wrongly convicted in cases involving undercover police officers should be referred to a separate panel for consideration. Some of the activists who have kindly given their time to this book have given evidence, but are still waiting to hear the outcome.

New security and counter terrorism laws were framed as preventative measures, continue to be iterated through media discourse on environmental activism, which reinforced the power relations by placing activists as outside, unacceptable and enabling the neoliberal approach to become dominant in environmental discourse. In other words, media discourse constructed a narrative around environmental activism that reinforced a neoliberal approach in environmental discourse.

This book began by outlining the climate science behind the headlines, and the motivation of individuals to move beyond clicktivism, slacktivism and the professional activism of NGOs. The environmental movement and environmental political discourse through the lens of media practices has seen a divergence of approaches aiming towards the same ends, but through different routes. The political divide has opens up between economics and climate justice. As Klein (2014) observes "Ever since this political divide opened up over climate change, a great deal of social science research has been devoted to pinpointing precisely how and why political beliefs are shaping attitudes toward global warming" (2014, p. 36).

Climate skeptics still retain a voice within the media, Lomborg (2008) is still questioning the science behind the headlines, and Delingpole (2012) along with the Heritage Foundation continue to call environmentalists "watermelon". There remains public challenges the climate science through editorials, emotive language, and as Klein (2014) says,

> As 97 percent of the world's climate scientists attest, the Heartlanders are completely wrong about the science. But when it comes to the political and economic consequences of those scientific findings, specifically the kind of deep changes required not just to our energy consumption but to the underlying logic of our liberalized and profit-seeking economy, they have their eyes wide open. The deniers get plenty of the details wrong (no, it's not a communist plot; authoritarian state socialism, as we will see, was terrible for the environment and brutally extractivist), but when it comes to the scope and depth of change required to avert catastrophe, they are right on the money. (2014, p. 43)

Protest groups, and single issues cases (Australia and UK's Knitting nanna and anti-fracking movements) are also engaging in media movement concepts to influence public understanding. Chapters Two, Five and Six examined how a media frame is set early in representation. Once set it's difficult to challenge and alters public perception. As Gitlin (1980) states that "while media framing almost certainly helps shape public consciousness, it does not follow that the media by themselves determine the intensity or duration of the frame left in the public mind" (1980, pp. 141–142). The results is that "the media agree on a frame, and to the extent that the issue lies outside the direct an experience of the audience, the media keep the power to push forward their frames as the salient ones, the ones that condition and limit public discussion. The power of the media frame to identify the issue in the first palace preserves for the framers an important power over the very terms of public life" (1980, pp. 141–142). The languages, frames, metaphors and order of discourse by the media frames the protester as absent from the public consciousness.

Tactical use of the Internet with social media, tablets and smartphones provided a new platform from which activists could challenge the negative framing from political and traditional media language and discourse. Early on the Indymedia website (1999) taking inspiration from the Mexican Zapatista use of listserv provided a perfect platform to organize the Seattle and global justice/anti-capitalist protests. These events shaped the global justice movement and informed many of their actions. The Zapatista movement gained global awareness through the internet (cf. Kingsnorth, 2003; Krøvel, 2011 in Chapter Six). Moreover, the internet enabled activists to retain their repertoires of protest and adapt old tactics and strategies to new techniques (see the discussion on the swarm/swoop in Chapter Six). Yet, an over-reliance on new social media has almost turned environmental activism backwards. Too much focus on media movement concept and getting hits and likes meant many actions were dismissed by the general public as a middle class hobby. Some protesters are aware too much focus on social media and media tactics, and too little on political strategies can be detrimental. So too can remaining within the middle classes. Activists choose to bypass traditional media are attempting to steer clear of representations of violence in what Graeber (2002) notes "in the corporate media, the word 'violent' is invoked as a kind of mantra—invariably repeated—whenever a large action takes place" (2002, p. 66). Bypassing corporate media became easier with globalization of the internet and global justice movements culminating the Seattle protest in 1999, were as Graeber (2002) goes on to say important as:

> In North America especially, this [global justice] is a movement about reinventing democracy. It is not opposed to organisation. It is about creating new forms of organisation. It is not lacking in ideology. Those new forms of organisation are its ideology … ultimately it aspires to reinvent daily life as whole. But unlike many other forms of radicalism, it has first organised itself in the political sphere. (Greaber, 2002, p. 70)

However, attempts at traditional media practices in controlling the public and political sphere were unsuccessful in creating new forms of organization. Gatekeeping, codes of conduct, on sites tour, no media zones, open hours and media escorts all were attempts at infusing the order of discourse, as was the complaint to the IPCC. The media tent, first outside the protest camps, and slowly integrated and then disbanded led to antagonism between journalists and activists, and activists and novice protesters. A risk emerges of a hierarchal semi-quasi-structure within protests that was transferred to other protests (such as Occupy LSX which also had a media team). The best way to mitigate risk is to push for heterotopias, that includes unitary and caution in openness. Protest camps can aspire to be a spatial zone in which to offer alternative ways of living and social relations. The disconnect between the alternative media tents, and traditional media journalists only led to limited coverage, and undermining the action with a headline such as the *Daily Mail* newspaper's "What do we want? A nice Cath Kidson Tent, When do we Want it, Now!" (Fryer, 2009).

Technological advances in communication should make it increasingly difficult for journalists to retain this position as websites, blogs, smartphones and the internet enable activists to bypass mainstream media. The internet also takes local protest onto the global stage to generate more interest and more individuals to challenge the advanced liberal government approach to environmental discourse. The internet, technology, smartphone, video and citizen journalism, all require a decent amount of digital literacy, affluence to afford the technology all points toward the problem of class. Class emerges very strongly in the discourse, serving as an indicator of the change in demographics of many UK protest groups (see Chapter One and Saunders, Grasso, Olcese, Rainsford, and Rootes, 2012; Saunders, 2013).

Activists DG believes that if protesters only engage with the left newspapers, such as *The Guardian*, the result is limited demographics to appeal to as "white liberal press and you have predominantly white people at the Camp" (2014). Guardian reading demographics that excludes raising awareness on class, digital divide low socio-economic communities which is problematic for any protester if there is to be genuine movement building.

To learn lessons from the different media strategies as contemporary activists have learnt from the past events and implemented new approaches (such as rotation of people to avoid one person becoming a figurehead, ch 5). Social media has provided the platform to activists to contest mainstream journalism. Activists realize that social media needs to be supplemented with the face-to-face practice of protest camps. The horizontal radical environmental politics enable anyone with a website or smart phone to challenge mainstream journalists. Although as the activists note this often only results in preaching to the converted. So We Stand activists NP notes:

> I think it's about having a critical relationship with it, understanding it and knowing its strengths and its pitfalls and who the audience is and who the audience isn't … it's about understanding where the overlaps can come and how you can use both situations there to your advantage. (Interview with NR, 2011)

Activist NR's observations are echoed by Richard, in that "there needs to be a constant awareness that "there's a permanent of back to nature kind of thing which environmental activists used to get into" (interview with Richard Herring), although as discussion in Chapter Six shows, today's activists have a strong focus on social media in the move to becoming a media movement.

Space, Power and Protest

Feigenbaum, Frenzel, and McCurdy (2013) note that "what makes protest camps distinguishable from other modes of protest is largely their attempt to create sustainable (if ephemeral) structures for ongoing protest and daily living" (2013, p. 30). The protest camps are a mirror held up against as an alternative, climate justice solution to climate change. Most camps have composting toilets, communal cooking, communal living, consensus decision-making, and wind and solar power. When the protest camps remove themselves from heterotopias, then they begin to have a relationship with mainstream party politics. In conforming through media discourse to create a hierarchy of working groups, spokesperson, and movement building then the protest camp begins to engage with party politics. The creation of media tent and fences means the protest camp is no longer an alternative space, but a space of counter-politics. To paraphrase Death's (2010) interpretation of Foucault—there is now a desire to be governed not like that, but in terms of another form of counter–conduct. The protest camp shifts from an alternative space to one that wants to be governed by a top-down hierarchical system, led by the media team, which forms its own eco-governmentality but one founded on civil disobedience and direct action. Activists can invert hierarchical power relations between themselves and journalists by protesting at the site of contestation, away from the media eye. Through their social practices (liminoid practices) activists create not a counter–space, but an alternative, which demonstrates a way of being ecologically aware. However, as this book has discussed, environmental activism to survive and challenge the dominant discourse it must retain the repertoires of protest within the heterotopia of the camps and actions.

Once a protest collective or group remove or alter the liminoid practices, loose focus or objectives or agitate the inside/outside dichotomy, they also remove the ability to challenge advanced liberal governments in environmental activism. Once activists look to be conducted through counter-conducts (cf. Death, 2010; Chapter Seven) they are no longer an alternative space to explore other options in environmental discourse, but conformed and contained through state apparatus of surveillance and fences. Remove the liminoid practice and there's no longer an inversion or subversion of traditional media practice, but conforming. Protest becomes formulaic, open to manipulation, a festival, and looses both its historical understanding and objective. Once the focus, and inside/outside dichotomy then the protest without assistance from the police or state will ceases, implode and loose the power base it has built up through liminoid practices, and media movement strategies. There are numerous reasons why, burn out, police pressure, state pressure, court case after court case, and so on, but every time a protest reaches its determinant function, and then loses it, it takes a good few years to "metamorphosis" into a new mobilization, and this action is at the moment repeated over and over again.

This book has examined the politics of protest between ecoActivist and governments through a media lens discourse. The research shows that the mainstream media frame environmental activism as part of environmental discourse, but excluded from mainstream party politics. Knowledge of environmental activism as part of environmental discourse, is formed through language to provide meaning, or maps of meaning as Hall defines discourse (see Chapter Two). In other words how we generate knowledge over time about a specific subject/discourse. In relation to the mainstream media and environmental activists, this book has argued that the relationship between political and environmental activists discourse has influenced the various ways the media frame environmental activism. The effect on public understanding of any protest, protest camp or mobilization sits in media frame that shapes maps of meaning and cultural codes.

In Chapter Three a discussion on political discourse of the Conservative political parties, both Margaret Thatcher's administration (1979–1992) and David Cameron (2010–present) co-opted environmental discourse into party politics, reveals changes in how journalists frame environmental activism. With the increasing administration of people (biopower) through policies of global, national and local environmental governance, and their associated techniques of self, journalists slowly silence the activist voice. Discursive struggles over the hegemonic position of environmental discourse became more apparent, tensions between activists and the state (such as the anti-roads and CJB) was reflected in the reporting of activism.

In contrast to Margaret Thatcher's preference for war over the environmental discourse (see Chapter Three), media coverage of the Conservative party election campaign, Vote Blue, Go Green (2006) acknowledges the shift to "green" as an electioneering tactic. *The Sunday Times*' headline, Cameron goes in search of green credentials at Ice Station Dave (White, 2006); whilst the *London Evening Standard* and The *Sun* both saw the shift towards environmentalism as a political move. The *London Evening Standard* defined the move as "Brown and Cameron vie for green vote" (Waugh, 2006) and *The Sun* "Green for Go, Dave" (Anon, 2006).

The development of party political discourse into environmental discourse led to struggles over the best solutions to climate change. The co-opting of environmental discourse by mainstream parties increasingly became the dominant position within environmental discourse. The media's support for the co-opting of environmental discourse was reiterated with the emergence of global environmental governance. The move by the Conservative party to Vote Blue, Go Green, meant most journalists supported the move, in the context of a global shift (such as the Kyoto Protocol) towards economic led solutions became the central tenet of environmental discourse. Moreover, there was no evidence of activists' engagement with the reworked green capitalism of environmental discourse. The media reports of the Vote Blue, Go Green Campaign to the exclusion of NGO such as Greenpeace or Friends of the Earth; and no reference to ecoActivism.

Revisiting and Looking Forward

Ever since Alison Anderson wrote *Media, Culture and the Environment (1997)* the first in-depth analysis of how journalists report on environmental discourse, the field has grown exponentially. Lester's work (2007 and 2010), and Simon Cottle's work provides insight into the practice of environmental activism and media discourse in Tasmania, to show how Australian environmental activist collectives relate with the media. The book has shown how the relationship between environmental activists and mainstream politics is defined by discursive struggles. Environmental activism began in the dominant role of environmental discourse, only to be challenged by the neoliberal policy political economics. The shift towards new environmentally focused markets, juxtaposed with the application of biopower frames environmental activists as abnormal. Moreover, a consequence of this is that it depoliticizes any environmental activist's demands, whilst persuading the individual to consume their way out of environmental problems. The work reported here also interprets original data and adding to public understanding of how an individual can when equipped with the right information contest political decision makers that effect their lives, to take a positions on climate changes, by inverting and subverting traditional media practices.

This book has also attempted to give a voice to individuals who for some have been bringing together different collectives, sharing knowledge and liminoid practices, teaching the next generation and passing on knowledge from lessons learnt. Stalwarts John Stewart reiterates Hamish idea of developing better tools, if ecoActivism is to retain a presence in environmental political discourses.

> to get across to ordinary householders was that you can make a difference because most people think they can't make a difference with something as big as that, and so I that sense it's hope that, not that we would do it for them but that they could be part of something. They should help to change things ... hope that they can take action because if people feel no hope at all, then actually, they feel, "Well, what's the point?" You've got to give them hope that they can make a difference. (2011)

Activists understand that media and direct action strategies that rely on Facebook and Twitter are often naive and vulnerable. Social media is useful in challenging the dominant positions, as shown in Glass's observations on the ability to get media images out that challenge a dominant position. However, they also realize that for the movement to survive they need to combine the tactics of social media with face-to-face planning of direct action to ensure objectives aren't lost in the media frenzy.

For a period, 1997–2011, traditional media coverage followed a similar pattern encouraged by laws and surveillance techniques that criminalize elements of protest. The range of environmental protest over the past 40 years, since the Earthrise image and the Physlis McCormack sailing towards the Amitcka archipelago, then through the 1990s, environmental activism from single issue (road building) to wider global movements, and onto global justice, and today the battle continues with (such as fracking) each iteration a new political mobilizations taking with it a history and knowledge for all the previous moments. Each moment, each movement, each mobilization add and takes away something from the movement, but despite the laws, a very different world to more than 40 years ago and the fledgling beginning of protest movements, civil liberties, equalities, social justice, and the environmental continues today. So battles have been won, some lost, and some are just having a quiet sit down. As John Dryzek says, "the storyline of green radicalism points to multi-faceted social and ecological crisis which can only be resolved through radical political action and structural change" (1997, p. 185).

A Final Word

I'll leave the last words to two activists, Stalwart Peter Willcox who famously was captain of the first Greenpeace Rainbow Warrior ship that was blown up by French secret police, in New Zealand (1985); and Australian tree dweller, Miranda Gibson. Miranda spent 451 days sitting in an almost 200 foot (60 metres) old Eucalyptus tree in Tasmania's southern forest (2014). The area was earmarked for logging by Malaysian Company, Ta Ann. When both were asked what continues to motivated them, Willcox an activists for over 40 years, and Miranda a school teacher prior to her action says: her motivation to stay in the tree tops for so long was about "making a contribution, standing up for your rights, they may be old hippy mantras, but they remain at the heart of many environmentalists beliefs" (interview, 2015)

Miranda adds how she's hopeful her own experience made a political difference, she says

> my tree sit exposed both the unique beauty of our island state [Tasmania], and the destruction that is continuing in our forests. I believe my action, along with the actions of countless grassroots forest advocates … contributed to the government's decision to submit a World Heritage nomination early this year. (Interview with Gibson, 2015)

Miranda adds that whilst living on a platform in the tress she remained connected to the world via social media and the internet, she says, "while in the tree I received messages from people from all walks of life and all across the globe from people inspired by my actions to get out and do something positive for their community and the environment" (Interview with Gibson, 2015).

Captain Peter Willcox has a similar outlook, as he says;

> I think it's really easy these days to feel crushed an impotent by society and government. If you can do something that makes you feel involved, makes you feel like you're making a contribution, I think it's the keystone for a much healthier and happier lifestyle. (Interview with Willcox, 2013)

Protest and ecoActivism is still thriving today, and not always in the media eye. Despite democratic countries around the globe continuing to amend public order and terrorism laws, increase state power to gather metadata there remains ecoActivists of all ages taking direct action, in the name of environmental justice for future generations. Luke's prediction in 1999 has come to fruition but ecoActivism continues to strenghten as new generations learn and grow the movements.

This book is in part a homage to those I have met and continue to see on the streets, in social spaces, sheds, boats, communes, vans, beaches—wherever there are people who see environmental injustice. The second set of people this book was written for are those seeking to challenge the idea a high carbon economy is the only economy. By understanding the history of ecoActivism plans can be made for the future. See you on the streets.

References

Anon. (2006) Green for Go, Dave, *The Sun*, April 18, p. 6.

Death, C. (2010). *Governing Sustainable Development: Partnerships, Protests and Power at the World Summit*, Hoboken, NJ: Taylor & Francis.

Delingpole, J. (2012). *Watermelons: How Environmentalists are Killing the Planet, Destroying the Economy and Stealing Your Children's Future*, London: Biteback.

Dryzek, J. (1997). *The Politics of the Earth: Environmental Discourses*, Oxford: Oxford University Press.

Fairclough, N. (1995). *Media Discourse*, London: Edward Arnold.

Fairclough, N. (2000). *New Labour, New Language?*, London: Routledge.

Frenzel, F., Feigenbaum, A., and McCurdy, P. (2014). "Protest Camps: An Emerging Field of Social Movement Research", *The Sociological Review*, 62(3), pp. 457–474. doi:10.1111/1467-954X.12111

Fryer, J. (2009). "What Do We Want? A Nice Cath Kidston Tent. When Do We Want It? Now!", *Daily Mail* (London edn.), 28 August, p. 22.

Foucault, M. (1972). *The Archaeology of Knowledge*, London: Tavistock Publishers.

Foucault, M. (1986). *The Care of the Self: The History of Sexuality*. Vol. 3, translated by Robert Hurley, New York, NY: Random House.

Foucault, M. (1991). "Governmentality", in Burchell *et al.* (eds.), *The Foucault Effect*, Hemel Hempstead: Harvester Wheatsheaf, pp. 87–104.

Foucault, M. (2008). *The Birth of Biopolitics. Lectures at the Collège de France, 1978–79*, edited by M. Senellart and translated by G. Burchell, Basingstoke: Palgrave Macmillan.

Gitlin, T. (1980). *The Whole Worlds Watching: Mass Media in the Making & Unmaking of the New Left*, Berkeley, CA: University of California Press.

Graeber, D. (2002). "The New Anarchists," *New Left Review* (13). Available at http://newleftreview.org/II/13/david-graeber-the-new-anarchists (accessed December 2012).

Great Britain. *The Climate Change Act (2007–8)*, London. Available at http://www.publications.parliament.uk/pa/ld200708/ldhansrd/text/81126-0006.htm#08112648000004 (accessed June 2010).

Great Britain. *Serious Organized Crime and Police Act, 2005. Chapter 15*. London: Legislation.gov.uk (accessed June 2004).

Klein, N. (2014). *This Changes Everything: Capitalism vs. the Climate*, London: Allen Lane.

Lester, L. (2007). *Giving Ground: Media and Environmental Conflict in Tasmania*, Tasmania: Quintus Publishing, UTAS.

Lester, L. (2010). *Media and Environment: Conflict, Politics and the News*, Cambridge: Polity.

Lomborg, B. (2008). *Cool It: The Skeptical Environmentalist's Guide to Global Warming*, London: Marshall Cavendish.

Kingsnorth, P. (2003). *One No, Many Yeses: A Journey to the Heart of the Global Resistance Movement*, New York, NY: Free Press.

Krøvel, R. (2011). "The War in Chiapas: The Fall and Rise of Independent Journalism", in A. Charles and G. Stewart (eds.), *The End of Journalism: News in the Twenty-First Century*, Oxford: Peter Lang, pp. 129–138.

Luke, T. (1999). "Environmentality as Green Governmentality", in E. Darier (ed.), *Discourses of the Environment*, Oxford: Blackwell, pp. 121–151.

Saunders, C., Grasso, M., Olcese, C., Rainsford, E., and Rootes, C. (2012). "Explaining Differential Protest Participation: Novices, Returners, Repeaters, and Stalwarts", *Mobilization*, 17(3), 263–280.

Saunders, C. (2013). *Environmental Networks and Social Movement Theory*, Bern: Bloomsbury Academic.

Thatcher, M. (1988). Margaret Thatcher's public speech to the Royal Society, Fishmongers Hall, City of London, 27th September 1988. Available at http://www.margaretthatcher.org/speeches/displaydocument.asp?docid=107346 (accessed 30 September 2009).

Oels, A. (2005). "Climate Change Governable: From Biopower to Advanced Liberal Government", *Journal of Environmental Policy and Planning*, 7(3), pp. 185–207.

Waugh, P. (2006) Brown and Cameron vie for green vote, *London Evening Standard*, April 20, p. 4.

White, R. (2006) "Cameron goes in search of green credentials at Ice Station Dave", *The Sunday Times* (London), April 2, p. 17.

INDEX

A

B

C

D

E

F

G

R

S

T

U

V

W

Z

www.ingramcontent.com/pod-product-compliance
Ingram Content Group UK Ltd.
Pitfield, Milton Keynes, MK11 3LW, UK
UKHW021837100926
8556IPUK00004B/393

9 781433 131189